WHITEHEAD'S PHILOSOPHY OF SCIENCE
AND METAPHYSICS

WHITEHEAD'S PHILOSOPHY OF SCIENCE AND METAPHYSICS

AN INTRODUCTION TO HIS THOUGHT

by

WOLFE MAYS

MARTINUS NIJHOFF / THE HAGUE / 1977

FOR LAURENCE

ISBN-13: 978-90-247-1979-2   e-ISBN-13: 978-94-010-1085-6
DOI: 10.1007/978-94-010-1085-6

CONTENTS

Early years. School at Sherborne and then Scholar and Fellow of Trinity College, Cambridge. Mathematical Writings: *A Treatise on Universal Algebra* (1898). Relations with Bertrand Russell, first as his teacher, later as collaborator in the writing of *Principia Mathematica* (1910-13). *London Period*: Professor at Imperial College. Contributions to the Aristotelian Society. Philosophy of Nature writings – interest in relativity theory. *American Period*: Professor at Harvard. Philosophical writings: *Process and Reality* (1929). C.D. Broad and Bertrand Russell on Whitehead.

On interpreting Whitehead. Two schools of Whitehead interpretation: (a) that there is a continuity between his earlier philosophy of science and later metaphysics, and (b) that the problems and issues he is concerned with in his later work are largely of humanistic origin. Reasons for believing that philosophical notions occur in Whitehead's early writings and scientific ones in his later metaphysics. *Whitehead and Language*. Difficulties in understanding Whitehead's language in his later work. Similarities between Whitehead's philosophy and Husserlian phenomenology.

PART II: PHILOSOPHY OF SCIENCE

Introduction to philosophical notions of *A Treatise of Universal Algebra*. The ideas of equivalence and identity. The nature of a calculus. Distinction between equivalence and identity. Replacement of the general notion of equality by that of matching. Theory of spatial congruence and measurement as based on notion of matching. Logical and mathematical reasoning as involving synthetic judgments. Substitutive schemes and symbolism. Quine's critique of Whitehead's idea of equality as involving confusion of sign and object. Similarity of Whitehead's views on mathematical reasoning with that of the intuitionists.

Construction of alternative models of the physical world in Whitehead's *Royal Society* memoir (1906). Whitehead's geometrical and physical ideas combined there with his studies in mathematical logic. Resemblance to Leibniz's doctrine of possible worlds. Five concepts or models of the physical world developed. Classical Newtonian concept, assumes points of space, matter and instants of time. Whitehead's objection that it bifurcates the world into separate classes of elements. Alternative (Leibnizian) concept involving linear entities and unifying space and matter. Definition of points in terms of class of linear entities. Relevance of memoir to Whitehead's later philosophy.

Whitehead's philosophy of nature as developed in his *Enquiry* (1919) and *Concept of Nature* (1920): concerned there with question of relating abstract notions of science to the data of sense-perception. Materialism and the bifurcation of nature into two systems of reality: (1) apparent nature and (2) causal nature. *Whitehead's theory of perception:* basic elements discriminated in sense-perception (a) events (b) objects – relation between (a) and (b) as involving bodily event of the observer. *Types of object.* Events and the general part and whole relation of extension. *Method of Extensive Abstraction:* derivation of ideal points and instants. Criticism of method as circular. Russell's discussion and use of the method.

Discussion in *Science and the Modern World* (1926) of the historical development of materialism from Greek times to the present day. Fate in Greek tragedy and the order of nature. Influence of scholastic logic and medieval theology on the development of science. Historical revolt at the Renaissance in religion and science. Experimental method and induction. Inconsistency in 18th century European thought between acceptance of scientific materialism and a belief in man as a self-determining organism. Protest against this attitude shows itself in 19th century literature, particularly in poetry. Influence of doctrine of evolution on scientific thought. The realm of possibility.

Difficulty in reconciling description of experienced time with that given by science. Husserl and Merleau-Ponty on the phenomenology of time-consciousness. Comparison with Whitehead's position. Northrop and Einstein on scientific concepts as free creations of the mind. Whitehead's views on congruence as connected with our recognition of sameness or uniformity in nature. Poincaré's concept of the congruence relation as conventional. Northrop's and Grünbaum's criticisms of Whitehead's account of simultaneity and congruence. Grünbaum's belief that the metric of psychological time is based on that of physical time.

PART III: METAPHYSICS

Philosophical systems. limitations of categorial frameworks implicit in past cosmologies. Whitehead's attempt to construct an alternative scheme to harmonise with present day scientific knowledge and take account of the rich variety of human experience. Descriptive character of the categoreal notions constituting the philosophy of organism. *Speculative philosophy and its method.* Whitehead's doubts as to the possibility of our ever finally formulating definite metaphysical principles. Errors in past philosophical thinking. General verification of a philosophical system to be sought in its practical success.

The Theory of Prehensions: concept of prehension covers not only physical transmission, but also perception, cognition and judgment in man. *Positive prehensions* deal with endurance of physical objects, *negative prehensions* with alternative states of affairs. *Physical and conceptual prehensions* – spatio-temporal aspects of events and the sensory perspectives marking them out. Reality as made up of ordered events exhibiting regularities of pattern (societies). *Corpuscular societies* comprise physical objects analysable into simpler objects, molecules, electrons, etc. Higher grades of such societies include animals and other living organisms.

Atomicity and continuity in perception. Physical world as consisting of atomic entities but perceived as made up of common sense objects. *Theory of symbolic reference*: physical events and sensory experience assumed to manifest a common relational structure. Sensory perspectives taken as standing as symbols for the causal activities in the physical world. Bodily experience and perception. Difficulties in relating the two perceptual modes, *presentational immediacy* and *causal efficacy*. Critique of Whitehead's and Russell's view that physical structures are isomorphic with sensory structures.

Normal and illusory perceptions: (a) perceptual propositional prehensions refer to authentic (veridical) perceptions and unauthentic (illusory) perceptions; (b) imaginative propositional prehensions refer to mental states as occurring, for example, in memory and imagination. *Intellectual prehensions*: (1) conscious perceptions – i.e., immediate perceptual judgments, and (2) intuitive judgments as occurring in thought and imagination. Three types of empirical judgment. Distinction between propositions and judgments. Derivative judgments and deductive reasoning.

Whitehead's critique of Hume's analysis of the causal relationship as one of expectancy. Hume's and Kant's denial of direct causal determination stems from their deletion of the perceptual process of passage from experience. *Perceptual Causality*: Causal determination as a characteristic of a whole occurrence, from which the more refined causal notions of science are derived, Michotte's experimental work on perceptual causality – habit and expectancy not crucial factors in giving rise to causal impressions. Critical examination of Michotte's work from an analytical standpoint.

Chapter 13. RELIGION, DEITY AND THE ORDER OF NATURE   128

Whitehead's Natural Theology. Belief that he is putting forward a new kind of ontological argument for the existence of God. Whitehead's approach primarily descriptive, concerned with those general features of order in the universe which he believes give rise to our particular religious ideas, emotions and forms of behaviour. *Dipolar nature of God*: (a) Primordial Nature – the conceptual aspect of our experience of an order of nature; (b) Consequent Nature – the multiplicity of events or the qualitative content ordered by (a). *The Order of Nature*: Whitehead expresses this order more formally in terms of his concept of an Extensive Continuum. Whitehead's linking of mathematical concepts to value concepts.

PREFACE

In this book I have attempted to give an account of some of the most important of Whitehead's philosophical writings – his writings on the philosophy of science as well as his metaphysics. I have tried to show that although there are novelties in Whitehead's later philosophy there are also continuities with his earlier work in the philosophy of science. For a more detailed account of Whitehead's metaphysics, I would refer the reader to my book *The Philosophy of Whitehead* (The Muirhead Library of Philosophy), Allen and Unwin, London 1959 (Collier Books, New York 1962). On the whole I believe my view of Whitehead in that work, at least as far as his metaphysics is concerned, is not materially different from that held in the present one, although there are some differences in emphasis and interpretation.

I wish to thank the administrateur délégué of the *Revue Internationale de Philosophie*, Brussels; the publishers Allen and Unwin, London; and Springer-Verlag, Heidelberg, for kindly giving me permission to publish amended versions of the following papers which I originally published with them: "Whitehead and the Idea of Equivalence" *Revue Internationale de Philosophie*, No. 56-57, 1961, pp. 167-184; "The Relevance of "On Mathematical Concepts of the Material World" to Whitehead's Philosophy", in *The Relevance of Whitehead*. Ed. Ivor Leclerc (The Muirhead Library of Philosophy) Allen and Unwin 1961, pp. 235-260; "Whitehead and the Philosophy of Time". *Studium Generale*. Springer-Verlag 1970, pp. 509-524, and in *The Study of Time*. Eds. J.T. Fraser, F.C. Haber, G.H. Müller, Springer-Verlag 1972. These papers now appear in their revised form as Chapters 3, 4 and 7 of this book.

WOLFE MAYS

University of Manchester

ABBREVIATIONS

AI   *Adventures of Ideas*

CN   *The Concept of Nature*

MC   "On Mathematical Concepts of the Material World"

MT   *Modes of Thought*

NL   *Nature and Life*

PM   *Principia Mathematica*

PNK   *An Enquiry Concerning The Principles of Natural Knowledge*

P of R *The Principle of Relativity*

PR   *Process and Reality*

PART I

THE MAN AND HIS WORK

CHAPTER 1

LIFE

In his "Autobiographical Notes,"[1] Whitehead tells us that he was born on February 15th 1861 at Ramsgate in the Isle of Thanet, Kent. His father was at first a schoolmaster, but later became an Anglican clergyman and Vicar of St. Peter's Parish near Ramsgate. It was through Whitehead's father that Bertrand Russell first made his contact with Whitehead. Russell relates that as a boy he had been told that the earth was round – which he did not believe. His people thereupon called the Vicar of the Parish, who happened to be Whitehead's father, to persuade him otherwise. "Under clerical guidance," he says, "I adopted the orthodox view and began to dig a hole to the Antipodes."[2]

In 1875, at the age of 15, Whitehead was sent to school at Sherborne where he studied not only mathematics, but also classics; and in his final year he was Head of the School. In 1880 he went to Trinity College, Cambridge as a scholar, became a Fellow in 1885 and remained at the College until 1910, but continued as a Fellow until his death. He tells us that during his undergraduate period at Trinity all his studies were in mathematics, pure and applied. His extra-curricular education was mainly derived from discussions with friends, undergraduates or members of the staff. He notes that by the time he gained his Fellowship, he knew nearly by heart parts of Kant's *Critique of Pure Reason*, although he was early disenchanted with it. As far as Hegel was concerned he was, he says, put off him by reading some of his remarks on mathematics – which struck him as pure nonsense.

Whitehead's Fellowship dissertation was on Clerk Maxwell's theory of electromagnetism. Because of this Russell notes that Whitehead was always thought of at Cambridge as an applied mathematician. As a result his first book *A Treatise on Universal Algebra* (1898) was not perhaps as well appre-

---

[1] Alfred North Whitehead, "Autobiographical Notes," in *The Philosophy of Alfred North Whitehead*, ed. P.A. Schilpp, Northwestern University Press, 1941, p. 3.

[2] Bertrand Russell, *The Autobiography of Bertrand Russell 1872-1914*, Allen & Unwin, 1967, p. 29.

ciated as it might have been. In this work he developed Grassmann's Calculus of Extension, Sir W. Rowan Hamilton's Quaternions and Boole's Symbolic Logic. And one of its aims was the almost Leibnizian one of developing a calculus to facilitate reasoning in every region of thought. A second volume was projected by Whitehead and he worked on it for several years. But this work seems to have been overtaken by his studies for *Principia Mathematica*. In 1906, during the time he was collaborating with Russell on *Principia Mathematica*, Whitehead published "On Mathematical Concepts of the Material World" in *The Philosophical Transactions of the Royal Society*. He endeavoured to show how alternative concepts of the material world (or cosmologies) could be constructed. He elaborates five alternative concepts and a number of sub-concepts, and also sketches the beginnings of what was later to become the *Method of Extensive Abstraction*. This paper was perhaps the first attempt to apply mathematical logic to a concrete subject matter – in this case physics. At about this time he also produced two small monographs on the foundations of geometry. *The Axioms of Projective Geometry* (1906) and *The Axioms of Descriptive Geometry* (1907).

Principia Mathematica (3 vols., 1910-13) was written in collaboration with Russell, who had already written *The Principles of Mathematics* (1903). Russell had been one of Whitehead's students at Trinity, and his Fellowship dissertation, later published as *Foundations of Geometry* (1897), had been influenced by Whitehead. The primary aim of *Principia Mathematica* was to show that all pure mathematics follows from purely logical premises and uses only concepts definable in logical terms. There was to be a fourth volume on geometry written by Whitehead himself. It seems probable that some parts of this projected volume were incorporated in "On Mathematical Concepts of the Material World" and in the more technical sections of his philosophy of nature writings, where he was trying to ground geometry on the crude data of sense-perception. In this fourth volume, according to Russell, Whitehead proposed "to treat a space as a field of a single triadic, tetradic or pentadic relation, a treatment to which he said he had been led by reading Veblen."[3] This treatment would seem to resemble Whitehead's account of the derivation of geometry in "On Mathematical Concepts of the Material World." On the other hand, Russell has also told us that he owed to Whitehead the suggestion for treating points, instants and things of the physical world as constructions, and that what he has to say on these topics "is, in fact, a rough preliminary account of the more precise results which he [Whitehead] is giving in the fourth volume of our

[3] Bertrand Russell, "Whitehead and *Principia Mathematica*," *Mind*, Vol. LVII, 1948, p. 138.

Principia Mathematica."[4] From this it would appear that the fourth volume would have contained the results of his thinking on geometry as a physical science, results which one finds in his philosophy of nature writings. But these two accounts do not conflict, since Whitehead's later treatment of points grew out of this earlier work. However, there seems little liklihood of our ever finding out what exactly the contents of volume four were to be, since all Whitehead's unpublished manuscripts were destroyed at his death.

Principia Mathematica was undoubtedly one of the most important contributions in the field of mathematical logic and the foundations of mathematics of this century, and the part played by Whitehead has sometimes been underestimated. Russell on a number of occasions has tried to set the record right. He points out that throughout the years 1900 to 1910 both he and Whitehead gave the bulk of their time to the writing of *Principia Mathematica*. The philosophical problems were largely left to Russell and Whitehead concerned himself in the main with the mathematical ones. However, Russell says, this only applies to first drafts: "When one of us had produced a first draft, he would send it to the other, who would modify it considerably. After which, the one who had made the first draft would put it into final form. There is hardly a line in all the three volumes which is not a joint product."[5]

In 1911 Whitehead published his little book *An Introduction to Mathematics*. This work shows that Whitehead could write clearly and simply about technical questions, although in his later writings he became a somewhat obscure and careless writer. He retired from his Cambridge teaching post as a Senior Lecturer in 1910. He then held various appointments at University College, London from 1911-14, and was Professor of Applied Mathematics at the Imperial College of Science and Technology from 1914-24. During this period he became interested in educational administration (he was Dean of the Faculty of Science in the University for a time) and the philosophy of education. The essays written over this period occur in *The Organisation of Thought* (1917), republished as *The Aims of Education* (1929). Towards the latter part of this period he began working on epistemological problems and also took part in the discussions of the Aristotelian Society. He was much concerned with bridging the gap between physics and sense-perception. These topics were discussed in his Aristotelian Society papers and published in the Society's *Proceedings*. The most important books published in this period were his *An Enquiry Concerning the Principles of*

[4] Bertrand Russell, *Our Knowledge of the External World*, Allen & Unwin, 1949 reprint, p. 8.

[5] Bertrand Russell, *My Philosophical Development*, Allen & Unwin, 1959, p. 74.

Natural Knowledge (1919), *The Concept of Nature* (1920) and *The Principle of Relativity* (1922).

In these works Whitehead concerned himself with the philosophy of mathematical physics at a time when relativity theory was influencing discussions of the foundations of the subject. In *An Enquiry Concerning the Principles of Natural Knowledge* and *The Concept of Nature*, he was largely concerned to overcome the dualism between a world of objective scientific objects and a world of private and mind-dependent sense-qualities, and to show how the refined mathematical concepts of physics were rooted in our actual sense-experience. He believed the world described by the physicist really did exist, though perhaps not quite in the neat form in which it appeared in the physicist's equations. It is here that his *Method of Extensive Abstraction* proved its worth, since as a logical instrument it enabled him to show the connection between such geometrical concepts as volumes, straight lines, points, and our sense-perceptions.

In *The Principle of Relativity* Whitehead concerns himself with the general theory of relativity, and among other things he there suggests a new law of gravitation from which certain empirically testable results might be deduced. Whitehead's theory starts from somewhat different assumptions from those of orthodox relativity theory. He arrives at the transformation equations of the special theory of relativity from very general considerations, without reference to empirical facts about the velocity of light and the synchronisation of clocks by light signals. Whitehead also claimed that space-time is uniform as opposed to the orthodox relativist position that the structure of space-time is not uniform but varies with its contents.[6] In recent years some attempt has been made to rehabilitate Whitehead's theory by John L. Synge and others, but this attempt has not been entirely successful.

Whitehead was invited to join the Faculty of Harvard University as Professor of Philosophy in 1924. He was then 63 and as Broad remarked: "It has been given to few men to start a new career in a new country at the age of 63 and thereafter to spend at least twenty more years in extremely original intellectual activity of the highest quality along quite fresh lines. Whitehead's output during this last period would be astonishing even in a man still in the prime of life."[7] The offer from Harvard was of a five year appointment. After two years of teaching he was asked to stay on as long as he could – 13 years in all. At his retirement in 1937 as Emeritus Professor he became a Senior Fellow of Harvard.

[6] See C.D. Broad's excellent short summary of this work in "Alfred North Whitehead (1861-1947)," *Mind*, Vol. LVII, 1948, pp. 142-144.

[7] *Ibid.*, p. 140.

During his American period his writings were primarily philosophical and took on a more metaphysical character. This showed itself in the books he wrote during that period, *Science and the Modern World* (1925), *Religion in the Making* (1926), *Symbolism its Meaning and Effect* (1927) and *The Function of Reason* (1929). In *Process and Reality: An Essay in Cosmology* (1929), Whitehead elaborated his philosophy of organism. This was first given as a series of Gifford lectures at Edinburgh University in 1927-28. Broad has said of *Process and Reality*, that "it is one of the most difficult philosophical books that exist, it can vie in this respect with the work of Plotinus and Hegel." Broad does perhaps exaggerate the difficulties of this work: some parts are clearly and intelligibly expressed – though other parts seem to exemplify Hegel's saying that "The Owl of Minerva flies only at night." *Adventures of Ideas* appeared in 1933, *Nature and Life* in 1934 and *Modes of Thought* in 1938. Whitehead also wrote and published a number of essays during this period. Some of them are reprinted in *Essays in Science and Philosophy* (1947). Lucien Price has recorded the conversations he had with Whitehead over a period of years during his Harvard years in *Dialogues of Alfred North Whitehead* (1954).

A Treatise on Universal Algebra led to Whitehead's election to the Royal Society in 1903. Nearly 30 years later he was elected to a Fellowship of the British Academy as a result of work in philosophy beginning in about 1918. He received the Order of Merit in 1945, and died in Cambridge, Mass. on December 30th 1947.

Broad mentions a very important fact about Whitehead, namely that at no period of his activity was he "philosophising on an empty stomach." Apart from his technical mastery of mathematics and symbolic logic, Whitehead had, he says, a wide and deep knowledge of history in general and of the history of natural science in particular. He was steeped in the best European literature and he had a sympathetic understanding of certain forms of religion. His personality was sane and balanced, free from crochets and grievances. "His work, therefore, never gives the impression of thinness and clever-silliness which sometimes characterises the productions of highly intelligent writers who have lacked these advantages."[8]

We have Russell's own tribute to Whitehead's personal qualities and his excellence as a teacher. Russell notes Whitehead's extraordinary capacity for work. He relates how he and a friend once went to visit Whitehead and watched him covering page after page with symbols. "He never saw us, and after a time we went away with a feeling of awe."[9] Socially, Russell goes on,

[8] *Ibid.*, p. 141.
[9] Bertrand Russell, *Autobiography*, p. 129.

he appeared kindly, rational and imperturbable and was certainly not that inhuman monster: the rational man. "Like other men who lead extremely disciplined lives, he was liable to distressing soliloquies, and when he thought he was alone, he would mutter abuse at himself for his supposed shortcomings."[10] We get an echo of this in Whitehead's "Autobiographical Notes": "The shortcomings of my published work, which of course are many, are due to myself alone."[11]

[10] *Ibid.*, p. 129.

[11] "Autobiographical Notes," p. 14. It is interesting to note here that two important philosophers who have had a considerable influence on logical thinking were both students of Whitehead: Russell at Cambridge and Quine at Harvard.

CHAPTER 2

GENERAL INTRODUCTION

I. ON INTERPRETING WHITEHEAD

In this account of Whitehead's thought I shall among other things be concerned with his mature philosophy as well as some of his earlier more mathematical and logical work. I believe that a study of Whitehead's pre-metaphysical writings can throw light on his later philosophy, and also show its relevance for modern thought. Too often Whitehead's views have been neglected because of his later reputation for obscurity in his metaphysical writings. Although few would wish to say that his major philosophical work *Process and Reality* is completely without philosophical significance, what this significance is, is not always clear to the cursory reader. Further, a good number of philosophers are put off by what looks like an attempt to inject "value" and "feeling" into physical nature.

The reader may well ask why should I study such a difficult and obscure philosophy as Whitehead's? To this one might reply that in the first place it is the work of a man who had distinctive and often original ideas in logic, mathematics and physics – he was after all the co-author of *PM* – and there is some reason for believing that they were carried over to his later philosophy. He was not a philosopher who philosophised, as Broad puts it, "on an empty stomach," and in his writings he touched upon many problems of vital intellectual interest.

To add to the reader's difficulties in understanding Whitehead's later philosophy there are, when it comes to interpreting Whitehead, at least two different schools of thought. We can on the one hand, try to show that the key notions in *PR* have some continuity with the concepts developed in his earlier studies in mathematics, logic and the philosophy of nature. On the other hand, we can take the view that there has been a radical shift of interest in the problems and issues Whitehead is concerned with: that they are now of a purely philosophical sort. Thus it is argued that Whitehead's later philosophy is not simply an extended version of the earlier philosophy of nature: his philosophy has been radically changed, since it now takes account of metaphysical problems which his early work explicitly did not. In constructing his metaphysical system, this argument goes on, Whitehead relied largely upon his humanistic reflections upon experience and his reading over the years in religion, history and literature. For this reason, it is said, Plato and Aristotle, the fathers of modern philosophy, are perhaps better guides for an understanding of Whitehead's later work than is his early philosophy of nature.

Whitehead's wide reading in the humanities is beyond doubt and his philosophy as may be seen, for example, in *SMW* is essentially a reaction against the older scientific materialism which expelled man from nature. But this does not mean that Whitehead is in any way anti-scientific and that the results of science have no place in his philosophy. He is rather arguing that science needs reinterpretation so as to take into account what he conceives to be a basic fact about nature: that it is essentially "holistic." It is also true that in his later writings he draws explicit comparisons with Plato, Descartes, Locke, Berkeley and Hume. Despite this one cannot overlook his trenchant criticisms of some aspects of Aristotle's thought, particularly the subject-attribute mode of expression which he felt had obstructed the development of philosophy. Whitehead would probably himself have suggested that some knowledge of Greek mathematics and science would enable one better to appreciate the philosophies of Plato and Aristotle. Because of all this it seems to me unlikely that Whitehead's later work breaks completely with his earlier philosophy, which among other things concerned itself with logic, mathematics and physical questions. In any case Whitehead would have considered mathematics to be as much a humanity as a science. As he says of it, "Pure mathematics in its modern developments, may claim to be the most original creation of the human spirit ... the pursuit of mathematics is a divine madness of the human spirit, a refuge from the goading urgency of contingent happenings."[1]

Though I readily admit that there are novel elements in *PR*, one can nevertheless find family resemblances between the topics discussed there and those discussed in his earlier philosophy of nature writings. Among these are: the emphasis on process, the relatedness of nature and the ingression of objects. I am not contending that one cannot understand Whitehead's metaphysics unless one has mastered *Principia Mathematica*. This would be an extreme caricature of my position. A knowledge of *Principia Mathematica* would no doubt be useful, but it is by no means essential. What I think is

---

[1] *SMW*, p. 32-33.

needed is some understanding of Whitehead's early logical work and his philosophy of nature, so as to avoid reading *PR* only within the context of traditional philosophy. I believe that with some such preparation, it should be possible to obtain an insight into some of the more difficult parts of *PR*.

As against those interpreters of Whitehead's work who make his scientific and metaphysical writings independent of each other Palter, for example, argues that such a clear division does not exist, as certain metaphysical considerations occur in Whitehead's earlier writings and certain scientific questions are of importance for an understanding of his later metaphysics.[2] It is therefore reasonable to assume that there is some overlap between the different periods of Whitehead's thought, and that his earlier more clearly expressed writings have some relevance for an understanding of his later more obscurely expressed ones. Palter also believes that Whitehead developed a metaphysical system to take account of the results of science and the data of human experience, and that he uses the phrase "metaphysical system" not so much in the traditional sense, but rather in the sense of a general language or group of ideas which could equally portray physical or psychological data.

II. WHITEHEAD AND LANGUAGE

One of the difficulties involved in understanding Whitehead's later writings and especially *PR*, is the somewhat esoteric terminology which he uses there, and which the reader has to master before he can obtain some understanding of this work. What I shall therefore try to do is to express Whitehead's ideas in fairly familiar language as far as this is possible, and where appropriate refer to his earlier writings, where often similar problems are discussed. It seems to me that if the reader can get behind Whitehead's language then he will find that his ideas are not without some philosophical significance. Broad has put it in this way, "I feel fairly certain that there is something important concealed beneath the portentious verbiage of the Gifford lectures."[3] However, whether you take the language used in *PR* as "portentious verbiage" or "technical vocabulary" will depend largely on your philosophical viewpoint. If like Broad you accept a common sense view of the world, "the natural attitude" as Husserl calls it, you will hold that philosophy will find the familiar terms of everyday speech adequate for its purpose. If, on the other hand, like Whitehead you wish to put aside the natural attitude,

---

[2] Robert M. Palter, *Whitehead's Philosophy of Science*, Chicago, 1960, cf. p. 1.
[3] C.D. Broad, "Alfred North Whitehead (1869-1947)," *Mind*, Vol. LVII, p. 145.

then you may need to introduce new technical terms to take account of novel ways of looking at experience.

Presumably this is Whitehead's reason for making use of a special technical vocabulary in *PR*. A feature of some of the terms he uses there is that they are applicable to both physical and psychological events. He often produces what in effect are new terms by extending the meaning of common words and phrases so as to give them a wider generality. Thus in borrowing from ordinary language such terms as "feeling," "concept" and "mental," he considerably extends their meaning. For example, when Whitehead talks of "conceptual feelings" he usually has in mind the notion of formal structure, whilst by "physical feelings" in their simplest form he refers to the transmission of physical energy in nature. If these phrases are interpreted purely psychologically, they take on a much too specialised character. We may be led to assume that Whitehead is primarily talking about conscious states of mind, whereas he is also referring to processes in the physical world. In addition to terms taken from ordinary language and given an extended meaning, there are the words and phrases which are peculiar to Whitehead's philosophy itself, such as prehension, actual occasion, eternal object, ingression, subjective aim, objective immortality, causal efficacy and presentational immediacy.

Whitehead's actual views on the nature of language are closely bound up with his views on perception. He believes that our common sense language only gives us an abstract picture of the dynamic world directly experienced by us, so that we come to conceive it as made up of independent substances characterised by qualities. Language, Whitehead points out, was designed to express such clear-cut concepts as "green-leaf," "round-ball."[4] He notes that such an observation statement as "This tree is brown," is highly elliptical and deletes the whole context of perception, namely, that I am in a particular place and time observing this particular tree which is brown in colour. But he readily admits that the abstract system of concepts embedded in ordinary language has proved itself of considerable value in enabling us to handle our common sense world.

Whitehead traces the excessive trust among philosophers in our common forms of language back to the Greeks. They thought, he says, and here he quotes J.S. Mill,[5] that the class of objects made by the popular phrases of their country, was natural and all others arbitrary and artificial. As a consequence they believed that by determining the notions attached to common

---

[4] *PR*, cf. p. 234.
[5] *PR*, cf. pp. 15-16.

language they would become acquainted with fact. This belief, Whitehead argues, vitiated the philosophy and physics of the Greeks as well as that of the Middle Ages.

III. WHITEHEAD'S PHILOSOPHY AND PHENOMENOLOGY

Whitehead's philosophy has, as has sometimes been remarked, a suprising resemblance to the doctrines developed by phenomenology and existentialism. This has been well brought out by Enzo Paci.[6] He tells us that both Whitehead and Husserl started life as mathematicians and were influenced by developments in logic, and both finally came to accept the view that the scientific world picture gives us an abstract one-sided account of things. They both argued that philosophers ought to get back to the rich variety of our direct human experience, putting on one side our everyday and scientific attitudes to the world. Further, Husserl's attempt to disentangle the complex interconnected set of concepts (or essences) discernible in our thoughts and perceptions – those of logic, of number, of qualities such as colour or musical pitch – bears an interesting resemblance to Whitehead's discussion of what he terms eternal objects.

Whitehead attacked scientific determinism, the philosophical expression of the mechanical world picture of seventeenth century physics, which led experienced nature to be denuded of its volitional, purposive and aesthetic properties. In this context Whitehead believed that by means of poetic language we can obtain some measure of insight into the real efficacious world of events, as in his writings the poet endeavours to express the more aesthetic and dynamic aspects of our experience.

Whitehead's belief that poetry can express essential aspects of our experience much more than can the abstractions of science, has also something in common with the approach of existentialists such as Sartre, who in their writings attempt to portray the human situation in all its richness, concreteness and subtlety. Whitehead denies that our perceptual world is simply made up of fragmentary sense-data devoid of all aesthetic and valuational characteristics. When we hate, he says, it is a man we hate, a causally efficacious man and not a collection of sense-data. He would emphasise that we have a direct awareness of the aesthetic qualities of things, for example, the pleasantness of a landscape or the threatening look of a storm cloud, and that these qualities are not simply projections of our minds.

---

[6] Enzo Paci, "Ueber einige Verwandtschaften zwischen der Philosophie Whiteheads und der Phänomenologie Husserls," *Revue internationale de philosophie*, XV, 1961, pp. 237-50.

We also find Whitehead criticising in much the same way as does the French phenomenologist Merleau-Ponty, the sharp division between body and mind introduced by Descartes, which he says has poisoned all subsequent philosophies. Like Merleau-Ponty, he brings out the important part played by bodily experience in perception, and argues that all our sense-perceptions – for example, the colours and sounds observed by us, have a bodily reference. Most of the current philosophical doctrines derived from Hume are defective, Whitehead says, by reason of their neglect of bodily dependency. He argues that our kinaesthetic and our emotional experiences are more basic than the visual perceptions from which physical science starts.

PART II

PHILOSOPHY OF SCIENCE

INTRODUCTION TO PART II

In this part I deal mainly with Whitehead's pre-metaphysical writings. I begin with an examination of some of the more philosophical parts of *A Treatise on Universal Algebra*, as they give a clear introduction to Whitehead's early thought. A good example of this is his discussion of the idea of equivalence. He also states succinctly his views on symbolism and mathematical reasoning, which for him, at least at this stage, has a synthetic rather than an analytic character.

I next examine Whitehead's Royal Society paper "On Mathematical Concepts of the Material World" (*MC*). Whitehead's aim there is to state axiomatically various possible ways of conceiving the nature of the physical world. Just as we can construct alternative geometries, so Whitehead believes one can construct alternative cosmologies. This memoir has a direct relevance to Whitehead's later writings. He criticises the classical Newtonian concept of the material world (or scientific materialism), since it splits nature into three exclusive classes: points of space, instants of time and particles of matter. One of his principal aims is therefore to reduce the number of relations involved in the classical concept. This he does by constructing alternative concepts which presuppose fewer entities and relations than the classical one. He also develops two theories of the derivation of points, one of which is the direct ancestor of the *Method of Extensive Abstraction*.

I then look at his philosophy of nature writings, particularly *PNK* and *CN*. Whitehead's starting point is our immediate sense-experience. By starting from the entities and relations given there, he hopes to be able to arrive at the precise concepts of physics, i.e., space, time and particles of matter. Whitehead points out that one of the consequences of classical materialism is the bifurcation of nature into two realities – an apparent nature and a causal nature. As in *MC*, he believes that we need to start from simpler and fewer notions, but he now starts from what he takes to be the simplest entities given in sense-experience – events involving passage and objects which endure. Whitehead also introduces his *Method of Extensive Abstraction* to demonstrate how points of space and instants of time may be derived from experienced events.

In *MC* Whitehead attempted to consider from a logical point of view, the classical Newtonian account of nature accepted by the scientists of the seventeenth and eighteenth centuries, and gave a precise account of some alternative concepts of the material world. In *SMW* he states his reasons for finding the materialist concept inadequate, and this he does by tracing the development of materialism from Greek times to the present. He also formulates a new concept of nature – a philosophy of organism – which he considers to be more appropriate to present day science than the older materialism.

I finally discuss Whitehead's views on time, which play a prominant role in his philosophy of nature and also in his metaphysical writings, and note some similarities between his account of time and that given by phenomenologists. Whitehead is critical of the absolute and relative theories of time and the way the mathematical concept of time has become absorbed into our ordinary life and language. Some of the criticisms which have been made of Whitehead's concepts of simultaneity and congruence are also examined.

CHAPTER 3

THE IDEA OF EQUIVALENCE

I. THE NATURE OF EQUIVALENCE

One of the clearest introductions to Whitehead's thought is to be found in certain parts of *A Treatise on Universal Algebra*: ideas are formulated there which also occur in his later work. A good example of this is his discussion of the idea of equivalence.

Equality, Whitehead points out, is one of the simplest notions we can come across in mathematics, as whenever we put a group of things in a class we set up equivalences between them. In philosophical circles this concept has not aroused very much interest in recent years. In the last century, however, there was considerable discussion as to whether the notion of equivalence can be taken over from mathematics, where it mainly means quantitative equivalence, and applied to logic, where it refers to the qualitative equivalence of classes.[1] What we do get, however, in contemporary philosophical writings, are numerous discussions of the notion of identity. This would seem to arise partly from present-day nominalist tendencies in logic and epistemology, where identity is regarded as a useful concept enabling us to put some order into our apparently synonym-ridden language.[2]

Whitehead's views on "equality" and "identity" deserve to be better known, as they have a direct bearing on some of the central features of his philosophical thinking. Only Quine seems to have referred explicitly to them, and this in a rather critical fashion. Whitehead's account is to be found in *A Treatise on Universal Algebra*[3] and in *The Principle of Relativity*.[4] In the former it occurs in the context of his discussion of the nature of a calculus. In the latter it occurs as a preliminary to his discussion of geometrical congruence and measurement. What emerges from both accounts, is that it is

---

[1] As W.S. Jevons tried to do in his "substitution of similars."

[2] Identity and equivalence have the following formal properties: they are symmetrical, transitive and reflexive.

[3] *A Treatise on Universal Algebra*, Cambridge, 1898.

[4] *The Principle of Relativity*, Cambridge, 1922.

never bare or absolute identity that we have to deal with. The notion of equality is always relative to a context or purpose.

II. EQUIVALENCE IN AN ABSTRACT CALCULUS

Whitehead's account of the notion of equality is then closely bound up with his discussion of the meaning of the relation of equivalence within an abstract calculus. An examination of Whitehead's views relating to equality will therefore at the same time involve us in an examination of his discussion of an algebraic calculus: in this case, his "Universal Algebra." By a Universal Algebra he means a calculus which symbolises the general operations of addition and multiplication, and in terms of which the equations of the system are set up. We must not, however, expect to find a rigidly formalised system of the *Principia Mathematica* type. Whitehead is not primarily concerned with formalising the steps involved in setting up such a calculus. The calculus he develops is not therefore characterised in terms of rules of substitution, definitions, axioms and theorems, as would be the case in a modern work on the foundations of mathematics.

Another reason for this lack of formal rigour, is that Whitehead still leaves a place for intellectual activities in mathematics.[5] They have their part to play in (a) the setting up of a calculus, and in (b) the manipulation of the symbols according to certain fixed rules. Whitehead's account therefore differs from more modern treatments, where a meta-language, whose function it is to lay down the permissible operations to be performed upon these signs, has replaced intellectual activity.

Whitehead begins his account by explaining the nature of the symbols or signs used in his system, together with their rules of manipulation.[6] In mathematics, he points out, we are exclusively concerned with substitutive signs. Such signs stand as proxies for other things, as is the case, for example, with counters in a game. At the end of the game the counters lost or won may be interpreted in the form of money. Mathematical symbols thus have a somewhat different character from the words of our ordinary language, which may also express our feelings and volitions. In a mathematical calculus we use substitute signs and reason by means of them according to fixed rules. It is assumed Whitehead tells us, that after a series of operations,

---

[5] In this respect his account is reminiscent of W.E. Johnson's account of "The Logical Calculus," *Mind*, Vol. I, N.S. 1892.

[6] *Universal Algebra*, cf. pp. 3-5.

they will denote when interpreted a true proposition about the things they represent.[7]

Whitehead defines a calculus somewhat pragmatically as "The art of the manipulation of substitutive signs according to fixed rules, and of the deduction therefrom of true propositions."[8] He explains this in more detail when he says, "The mind has simply to attend to the rules for transformation and to use its experience and imagination to suggest likely methods of procedure."[9] The rest is merely the physical interchange of signs instead of thought about the originals. Although Whitehead emphasises the part played by the mechanical manipulation of symbols in a calculus, some room is still left for the imagination in selecting the derivative chains of deduction.

Nevertheless, as Whitehead is fully aware, the prime concern of such a calculus is to replace intuitive inference by external demonstration. He regards inference as being "an ideal combination or construction within the mind of the reasoner which results in the intuitive evidence of a new fact or relation between the data."[10] In replacing such an intuitive construction by a formal calculus, the process of combination is now externally performed on a two-dimensional array of marks on paper. When in a symbolic calculus we come to interpret the resultants of our operations they are found to mean, Whitehead says, "the fact which would have been intuitively evident in the process of inference."[11] The whole of mathematics therefore consists for him in the organisation of a series of symbolic aids in the process of reasoning.[12]

III. EQUIVALENCE IN "A TREATISE ON UNIVERSAL ALGEBRA"

In Whitehead's account of a calculus in *A Treatise on Universal Algebra*, propositions take the form of assertions of equivalence. He uses the sign "=" to indicate that the sign or groups of signs on either side of it are equivalent, and may be substituted for each other. As Whitehead puts it, "One thing or fact, which may be complex and involve an interrelated group of things or a succession of facts, is asserted to be equivalent in some sense or other to another thing or fact."[13] It might seem from this that he regards

---

[7] *Ibid.*, p. 41.
[8] *Ibid.*, cf. p. 4.
[9] *Ibid.*, p. 10.
[10] *Ibid.*, p. 10.
[11] *Ibid.*, p. 10.
[12] *Ibid.*, cf. p. 12.
[13] *Ibid.*, p. 5.

equivalence as a relation between things rather than signs. However, he uses the term "thing" as a blanket-word to cover any object of thought or reality "concrete or abstract, material things or merely ideas of relations between other things."[14]

The relation of equivalence always occurs for Whitehead in connection with some definite field of enquiry or purpose. Two things are equivalent, we are told, when for some purpose they can be used indifferently, so that "the equivalence of distinct things implies a certain defined purpose in view, a certain limitation of thought or action."[15] Within this limited field no distinction of property exists between the two things. Equivalence as thus conceived is a relative notion; certain things may be equivalent for some definite purpose and not for another. We know that the way we arrange things in classes involves a certain arbitrariness. We could, for example, classify people according to their religion or age, etc. Two people x and y might then share the same characteristic of church-membership and yet differ from each other as far as age was concerned.

In explaining what he means by saying that two logical propositions are equivalent, Whitehead tells us that the equivalence is "expressed by $x = y$, when they are equivalent in validity." And by this is meant that "any motives (of those motives which are taken account of in the particular discourse) to assent, which on presentation to the mind induce assent to x, also necessarily induce assent to y and conversely."[16] We can only assume that by "motives to assent," Whitehead means those reasons, rational or otherwise, which persuade us to agree to the validity of both propositions. It is interesting to note how logical validity is connected here with intellectual feeling, and this would seem to link up with his much later account of judgment in *Process and Reality* in terms of the coherence of feelings.[17]

As the idea of equivalence is for Whitehead a relative conception, it requires special definition for any subject-matter to which it is applied. For example, in the context of the calculus developed in *A Treatise on Universal Algebra*, the definitions of the processes of addition and multiplication carry with them, he tells us, this required definition of equivalence.[18] Thus if a and a' be equivalent, they may both be used indifferently in any such series of operations of addition and multiplication. Whitehead's assertion that the equivalence relation always occurs in a specific context, bears some resem-

---

[14] *Ibid.*, p. 19.
[15] *Ibid.*, p. 5.
[16] *Ibid.*, p. 108.
[17] *Process and Reality*, cf. pp. 382-9.
[18] *Universal Algebra*, cf. p. 18.

blance to De Morgan's view that we usually think and argue in a limited universe, even when this is not expressly stated. Jevons, who followed De Morgan here, believed that all our identities are really limited to an implied sphere of meaning. According to him, when we say that " 'Mercury is a liquid metal', we must be understood to exclude the frozen condition to which it may be reduced in the Arctic regions."[19]

Whitehead makes a somewhat similar point when speaking of the ordinary propositions of geometry, which are the deductions of axiomatic geometry. He notes that the former are usually stated in an inaccurate abbreviated form. Thus, instead of our saying "Such and such axioms imply that the angles at the base of an isosceles triangle are equal," the protasis is in general omitted. The above statement therefore takes on the elliptical form, "the angles at the base of an isosceles triangle are equal."[20] In other words, there is a tendency in discussing formal relationships to overlook the fact that they depend for their meaning on the context in which they occur.

IV. TRUISM AND PARADOX

Whitehead carefully distinguishes the idea of equivalence from that of "mere" identity.[21] Equivalence, he tells us, implies non-identity as its general case. As we have seen, two things may be equivalent in certain respects, i.e., have a common property, although they may nevertheless differ from each other in other respects. Further, as far as identity is concerned, Whitehead conceives it as a special limiting case of equivalence.[22] It occurs for him largely in the arrangement of deductive proofs, where it is convenient to allow that a thing is equal to itself.

For Whitehead, all the things which for any purpose can be conceived as equivalent form the extension of some class. As an example, he considers the arithmetical equivalence $2 + 3 = 3 + 2$. This he takes to mean that the two methods of grouping the marks are equivalent, as far as the common fiveness of the sum on each side of the equation is concerned. But he goes on, "the order of the symbols is different in the two combinations, and this difference of order directs different processes of thought."[23] This difference of order is then regarded by Whitehead as being of some importance, since

[19] *Principles of Science*, Sec. Ed., p. 43.
[20] *The Axioms of Projective Geometry*, p. 2.
[21] *Universal Algebra*, cf. pp. 5-6.
[22] *Ibid.*, cf. p. 6.
[23] *Ibid.*, p. 6.

it refers to the temporal order in which the operations upon these symbols are to be performed.

Whitehead does not think it very helpful arithmetically, to define equivalent things as being merely different ways of thinking of the same thing as it exists in the external world. According to this view there is a certain aggregate of five things which we think of in different ways either as $2 + 3$ or $3 + 2$. As it is difficult to separate our modes of thinking from the things themselves, whether they be physical things or logical objects, Whitehead refuses to accept this distinction as a basis for mathematical reasoning.[24] The sharp distinction made by some present day philosophers between invariant logical objects and thought processes, is one which is alien to Whitehead at this stage.

Whitehead does not wish to identify the sign "$=$" as used in a calculus with the logical copula "is." As we have seen, two things b and b' are said to be equivalent when they both possess the attribute B, so that $b = B$ and $b' = B$, therefore $b = b'$. However, he argues, we cannot put this into the standard logical form b is B and b' is B, therefore b is b', without distorting the very nature of the equivalence relation.[25] For Whitehead, whereas the copula "is" states an identity between the two terms, an equivalence relation only states a partial identity.

Whitehead now gives an analysis of what is involved when we set up an equivalence relation. He points out that the equation $b = b'$ may be said to consist of two elements, which he calls the *truism* and the *paradox* respectively.[26] The *truism* is the partial identity of both b and b'; their common Bness. The *paradox* is the distinction between b and b', that they are two distinct things. In an equivalence relation then the two objects compared (x and y) possess a common property, but since they are two distinct individual things they must have in some degree diverse properties. Whitehead would then seem to be merely restating here Leibniz's principle of the identity of indiscernibles.

He goes on to point out that in a statement of equivalence, the *truism* is not usually alluded to, as the main stress is on the *paradox*.[27] For example, in the equation $2 + 3 = 3 + 2$ there is no explicit mention of the fact that both sides represent a common fiveness of number. The statement merely asserts that two different things $2 + 3$ and $3 + 2$, are in point of number equivalent. In order to verify this, we may simply note that the same factors,

[24] *Ibid.*, p. 6.
[25] *Ibid.*, cf. p. 6.
[26] *Ibid.*, cf. p. 7.
[27] *Ibid.*, cf. p. 7.

though in a different order, are mentioned on both sides of the equation. In this way we overlook the common property in terms of which they are said to be equivalent.

Whitehead makes the further point that in discussing the laws of a calculus, e.g., addition and multiplication, in terms of which our equivalences are set up, we stress the *truism*, i.e., the property owned in common by both sides of the equation. On the other hand, in the development of the consequences of a calculus we stress the *paradox*, namely the different propositions derived from the initial equations of our system.

V. EQUALITY IN "THE PRINCIPLE OF RELATIVITY"

Similar ideas are to be found in Whitehead's *The Principle of Relativity*.[28] He there discusses the notion of equality in connection with *congruence* or the quantitative equality of geometrical elements. Whitehead begins by considering Euclid's axiom that things that are equal to the same thing are equal to one another. But what, he asks, do we mean by saying that one thing is equal to another? If we mean equal in quantity (or magnitude) we must first know what we mean by quantity. However, if we define the quantity of a thing as its measurability in terms of a common unit, we are thrown back upon the equality of different examples of the same unit and thus beg the question.

To escape from this circularity, Whitehead widens his discussion so as to consider the notion of equality in its most general sense. Equality, he points out, would seem to have an obvious affinity with identity. Because of this, some philosophers would not, in considering the foundations of mathematics, want to draw any distinction between them. Although in certain usages of equality this may be the case, he does not, however, believe that this is the whole truth, as otherwise the greater part of mathematics would consist in tautologies.

Whitehead would nevertheless agree, that it is convenient for technical facility in the arrangement of deductive trains of reasoning to allow that a thing is equal to itself. However, he regards this to be merely a matter of arbitrary definition. He believes that the important use of equality occurs when there is a diversity of things related and an identity of character. In the light of the above it is difficult to see how Quine can assert, when speaking of

---

[28] Cf. Chap. III, "Equality." Whitehead tells us that a discussion of equality embraces in its scope congruence, quantity, measurement and identity.

A Treatise on Universal Algebra. "Whitehead's version of '=' as equivalence-in-diversity does not reappear in his later work."[29]

Whitehead's view that equality is an identity in difference is not peculiar to him, as it is already to be found in Bradley's *Principles of Logic.* Whenever we write "=" there must, Bradley says, be a difference, or we should be unable to distinguish the terms we deal with. He therefore refuses to accept the doctrine of simple identities or tautologies. To say that two things are precisely the same, we need, he points out, "first take A and B as differenced by place or time or some other particular, and then against that assert their identity."[30]

Whitehead believes that such an equivalence as $A = B$ is highly elliptical. There is always, he argues, an implicit reference to some character which both sides of the equation own in common. He brings this out in the following way. Let γ, he says, denote a class of characters, say colour. The equivalence (i) $A = B \rightarrow \gamma$, will then mean that the same member of the class γ, say c_1, qualifies both A and B, whilst the non-equivalence (ii) $A \neq B \rightarrow \gamma$ means that one member of the class γ, say c_1, qualifies A and another member, say c_2, qualifies B.[31]

Further, A and B may resemble each other in respect to one quality, but not in respect to another. For example, a polar bear and a zebra may possess the colour white, though the zebra may alone exemplify the colour black. Since two elements may resemble each other in some properties but not in others, there is thus a certain arbitrariness in the way we set up our equivalences. If we wished to arrive at an unambiguous statement of equivalence we would therefore have to restrict (i) above to mean that A and B each exemplify at the most one and the same character (and (ii) to mean that A and B do not possess the same one character). Thus it is only in such an ideal restricted situation that an identity relation can be said to exist between the terms related.

Whitehead concludes by suggesting that the general notion of equality as thus limited might be replaced by that of matching, in the sense in which colours match. Examples (i) and (ii) will then become respectively, A matches B in respect to the character γ and A does not match B in respect to the character γ. Starting from the notion of matching rather than equality, Whitehead proceeds to work out his theory of spatial congruence and

[29] "Whitehead and Modern Logic," in *The Philosophy of Alfred North Whitehead,* ed. P.A. Schilpp, p. 130.

[30] *Principles of Logic,* sec. ed., 1922, p. 374.

[31] Euclid's first axiom can then be stated in the form $A = B \rightarrow \gamma$ and $B = C \rightarrow \gamma$, implies that $A = C \rightarrow \gamma$.

measurement. As we shall see the view that "matching" is fundamental to the notion of equality is, as far as perception is concerned, already to be found in *A Treatise on Universal Algebra*.

VI. OPERATIONS, SYNTHESIS AND SUBSTITUTIVE SCHEMES

In *A Treatise on Universal Algebra* we are told that judgments can be founded on direct perception, as when we judge that two pieces of stuff match in colour.[32] Such judgments, he goes on, may also be founded on the respective derivations of the things judged to be equivalent. He includes under the heading of derivation physical and psychological derivations.[33] Two pieces of stuff may, for example, be judged to match in colour because they were dyed in the same dipping, or were cut from the same piece of stuff. The idea of derivation is, however, more general, since Whitehead also includes under it thought activities, as when, for example, we judge that two aggregates of three things and two things respectively make five things.

For Whitehead, logical and mathematical reasoning involves synthetic judgments. To illustrate this he asks us to consider an operation of synthesis "$\frown$" on two elements a and b, i.e., $\widehat{a\,b}$, which produces some third thing c, possessing the properties of both, thus $\widehat{a\,b} = c$. If a is 3 and b is 2, and the operation of synthesis is that of addition, then c will be 5.[34]

Whitehead's view that mathematical procedures have a synthetic character also appears in his later philosophy. We find him in *Modes of Thought* contradicting the fashionable notion that mathematical truths are tautologies. He refuses, for example, to regard the statement "twice three are six" as a tautology. This statement, he asserts, refers to a process and its issue: two such groups each characterised by triplicity are fused into a single group characterised by six.[35]

We have already noted that relations of equality always occur for Whitehead in some specific context, or as he puts it, in a scheme of things. By a scheme of things, he means any set of objects, denoted by a, a', etc., b, b',

[32] *Universal Algebra*, cf. p. 7.

[33] *Ibid.*, p. 7. Whitehead tells us, "The words operation, derivation and synthesis will be used to express the same general idea" (*ibid.*, p. 8).

[34] *Ibid.*, cf. p. 19.

[35] Cf. Lecture Five, "Forms of Process." Bradley says, *Principles of Logic*, p. 256. "Every inference combines two elements; it is in the first place a process, and in the second place a result. The process is an operation of synthesis; it takes its *data* and by ideal construction combines them into a whole."

etc., ... z, z', etc., which has some common or *determining* property, for example, colour possessed in different modes by different members of the set.[36] Any of these objects, for instance a, a', etc., are equivalent "=" if they possess the determining property in the same mode, e.g., if they are both coloured red. They are non-equivalent if they possess it in different modes, as is the case when one thing is red and the other blue.

In the case of two things with different defining properties, but with similar logical *characteristics*, a one-one correspondence can be established between them. Thus if a, b, etc., are equivalent in one scheme, then α, β, etc., will be equivalent in the other, and if m can be derived from a, b, etc., in one scheme, then μ will correspondingly be deducible from α, β, etc., in the other. All the theorems proved for one scheme can then be translated so as to apply to the other.[37]

In this way substitutive schemes can be set up exemplifying the same laws as the original more concrete scheme. Thus instead of reasoning about the properties of the original scheme, we can reason instead about the substitutive scheme and transpose back again at the end of our deductions. To illustrate this, he refers us to the theory of quantity, in which quantities are measured by their ratio to an arbitrarily assumed quantity, e.g., a metre rod.[38]

Further, instead of using substitutive schemes with naturally suitable qualities, we may, as in algebra, assign to arbitrary marks laws of equivalence identical with those of the original scheme. In this way we set up a purely conventional scheme. The value of such a scheme is that it enables us to a large extent to avoid the processes of intellectual inference altogether. As Whitehead puts it, "by the aid of symbolism, we can make transitions in reasoning almost mechanically by the eye, which otherwise would call into play the higher faculties of the brain."[39]

VII. MANIFOLDS

Whitehead's conception of a manifold plays an important part in his discussion of equivalence. By a manifold he means a conceptual property analysable into an aggregate of modes between which various relations may hold.[40] An example of a manifold would be the general property of colour

[36] *Universal Algebra*, p. 8.
[37] *Ibid.*, cf. p. 9.
[38] *Ibid.*, cf. p. 9.
[39] *Introduction to Mathematics*, p. 60.
[40] *Universal Algebra*, cf. p. 13.

which covers a large range of different colours; where colour is, in W.E. Johnson's terminology, the determinable and the different colours the determinates.

Another example is the idea of empty space referred to co-ordinate axes. Each point of space represents a special mode of the common property of spatiality. The fundamental properties of space are expressed in terms of these coordinates in the form of the geometrical axioms. The logical deductions from the axioms form the propositions of geometry and are investigated by a calculus; in this case axiomatic geometry. Whitehead identifies the properties of a positional (or abstract spatial) manifold, with the descriptive properties of a space of any arbitrary number of dimensions to the exclusion of all metrical properties such as distance.[41]

The concept of a manifold is then a very wide one, since it covers any conceptual property which may have a diversified set of interconnected determinates under it. It covers, for example, colours, musical notes, etc., as well as abstract spatial properties. Whitehead does not, however, make clear the precise ontological nature of such a manifold. He merely posits it together with its empirical exemplification in a concrete scheme of things.[42]

An interesting conclusion can be drawn from Whitehead's concept of a manifold, considered as a general property defining a range of more specific properties. The sum-total of such manifolds will give us something like a dictionary of all the possible properties, both spatial and qualitative, which could be exemplified in our experience (i.e. in a concrete scheme of things). Although Whitehead does not seem to spell this out in *UA*, if we were to consider manifolds in this more systematic fashion, we would arrive at something like a "realm of essences." Such a realm would resemble his much later conception of a "realm of eternal objects" developed in *SMW*.[43] In the "realm of eternal objects" Whitehead is also concerned with abstract properties such as extension and colour, which have the possibility of being exemplified in different modes in our experience.

VIII. CONTEMPORARY DISCUSSIONS

It is clear that Whitehead's conception of equality as an "identity in difference" is at variance with the views of those present day philosophers who accept a nominalist account of identity. According to them a statement of

---

[41] *Ibid.*, cf. p. 30.
[42] *Ibid.*, cf. p. 14.
[43] *Science and the Modern World.* Cf. Chapter on "Abstraction."

identity is to be regarded as made up of two different names each of which has the self-same object as its referent. Nevertheless, the difference between these names would seem to be more than a typographical one. Although the object referred to may be the same, we also deal with a difference in the meanings which these names express.

Quine, who seems to take up a nominalist position, is somewhat critical of Whitehead's ideas on equality. He tells us that some mathematicians have tried to look upon equations as relating numbers that are somehow equal but distinct. As an example of this, he quotes Whitehead's statement "that $2 + 3$ and $3 + 2$ are not identical; the order of the symbols is different in the two combinations, and this difference of order directs different processes of thought." It is debatable, he goes on, how much this defence depends on confusion of sign and object, and how much on a special doctrine that numbers are thought processes.[44]

In this paper on "Whitehead and Modern Logic,"[45] Quine examines Whitehead's views on equivalence in greater detail. Whitehead, he tells us, regarded "$=$" as expressing a relation of equivalence short of identity, as otherwise laws such as "$x + y = y + x$" would, like "$z + z$," make no assertion at all. This reasoning, he argues, loses its force if we distinguish closely between notation and subject-matter, i.e., the signs and the objects they refer to. Statements of identity that are true and not idle consist of unlike singular terms that refer to the same thing. The function of identity is then to bring together unlike terms which refer to the same thing.

Quine believes that we may show that "$x + y = y + x$" holds as a genuine identity, i.e., that the order of the summands is wholly immaterial, by superposing "$x + y$" and "$y + x$" in the manner of a monogram, so that we get a formula of the type "$z + z$." For reasons of economy, however, we impose instead an arbitrary linear notational order on the summands. One and the same sign is now expressed in two ways, "$x + y$ and $y + x$." The law "$x + y = y + x$" is then needed, Quine argues, as a means of neutralising the excess of notation over subject-matter.

For Quine then, the useful statements of identity are those in which the named objects are the same and the names different, as, e.g., in "Cicero is identical with Tully."[46] Statements of identity would, he thinks, be unnecessary if our language was a perfect copy of its subject-matter, and each thing had but one name. Indeed, for Quine the utility of our language lies in its very failure to copy reality in any one-thing-one-name manner.

[44] *Word and Object*, cf. p. 117.
[45] "Whitehead and Modern Logic," pp. 128-9.
[46] *Methods of Logic*, 1950, p. 209.

The weak point in Quine's argument would seem to be that he assumes that we know objects around us independently of any conceptual interpretation. He also seems to assume that our language possesses a store of precise synonyms upon which we can draw when we wish to refer to these objects. For Quine the excess of notation over subject-matter arises from the fact that we have a superfluity of names at our disposal to represent the things around us. However, what he overlooks is that this may really be due to our language referring not simply to things, but also to different characteristics or aspects of things.

There seems to be a tendency among some philosophers to talk about synonyms as if they really do in all cases express strict identities of meaning. However, to achieve such a precise synonymity would seem to be an ideal of thought, and would only be possible in a highly formalised artificial language. Synonyms usually possess different shades of meaning appropriate to the context in which they are used. By replacing one synonym by another in a sentence we may alter its whole character. Indeed it is hardly possible in our ordinary language, as opposed to a formal one, to find two words having in all respects the same meaning and being therefore interchangable.

In this connection, Bradley's criticism of the nominalist account of identity is worth mentioning. He tells us that the nominalist analysis of the statement "Dogs are mammals" assumes that "We have a set of individuals which in themselves are simply themselves. The difference asserted is the difference of the two signs 'mammal' and 'dog.'" On this view then, when we say "Dogs are mammals," we mean to assert that certain definite indivisible objects have got two names or have been christened twice, "and this is the real heart of your mystery."[47]

Bradley argues, however, that in the statement "Dogs are mammals," we deal with more than a duplicity of names. The terms "dog" and "mammals" refer rather to different attributes coexisting within the same thing. Although there are difficulties relating to the coexistence of these attributes, since the predicate term has a larger generality than the subject term, it is clear that an identity in ordinary language is not simply a synonym replacement relation. Bradley puts the nominalist position in its proper perspective when he says, "To those who believe assertions about things assert nothing but names, the universe has long ago given up its secrets, and given up everything."[48]

On a view like that of Quine's, the two sides of the equation $2 + 3 = 3 + 2$ are to be taken as simply two arbitrary names of the same thing,

[47] *Principles of Logic*, 1922, p. 177.
[48] *Ibid.*, p. 178.

namely 5, Superficially such a formal relation would seem, unlike one expressed in ordinary language, to exhibit precise synonyms. Quine tries to demonstrate this by introducing the alternative "$z = z$" notation. He overlooks, however, that the meaning of the monogram formula depends upon the fact that it stands as a proxy for the original linear formula. To this Quine might reply that the spatial order of the summands of the linear formula contains nothing of significance. To say, as Whitehead does, that it represents the temporal order of the intellectual operations to be performed upon the symbols, is for Quine to introduce irrelevant psychological considerations.

Whitehead is, however, not alone in emphasising that intellectual activities play a part in the construction of mathematical formulae, and that they are of some importance if we wish to understand the very nature of these formulae. The intuitionists, for example, conceive mathematics as a construction on the basis of the intuitively given natural numbers. We find Heyting interpreting the equation $2 + 2 = 3 + 1$ as an abbreviation for the statement: "I have effected the mental constructions indicated by '$2 + 2$' and by '$3 + 1$' and I have found that they lead to the same result."[49]

With regard to Quine's belief that the order of the symbols in a mathematical equation is entirely arbitrary, Whitehead has contended that a major characteristic of algebraic structures is that they attempt to exemplify significant patterns of order. In this respect they are different from the expressions of ordinary English, where the verbal order is in general irrelevant to the meaning of these expressions. Because of this, Whitehead would refuse to accept the view that our mathematical symbolism is purely arbitrary, having no connection with the ideas we wish to express.

---

[49] A. Heyting: *Intuitionism: An Introduction*, North Holland, p. 8. What is interesting about Whitehead's position is that although one of the proponents of the view in *PM* that mathematics is based on logic, he should yet have stressed the part played by intellectual activities in mathematical construction: a position one might expect from someone holding a synthetic view of mathematics rather than an analytic one.

CHAPTER 4

MATHEMATICAL CONCEPTS OF THE MATERIAL WORLD

I. THE IMPORTANCE OF THE MEMOIR

In 1906 there appeared in *The Philosophical Transactions of the Royal Society*, Whitehead's memoir "On Mathematical Concepts of the Material World"[1] (*MC*). Its object was to show how one could construct alternative models of the physical world. Although Whitehead had worked in pure mathematics and logic, his Fellowship dissertation at Trinity,[2] as we have seen, had been on Clerk Maxwell's theory of electromagnetism, in which he seems to have retained an interest throughout his life. In this memoir he brings together axiomatic theory and the physical view of nature in terms of the continuity assumed in electromagnetic theory. These two strands of thought are to be found running through his later writings, namely, an interest in abstract structure and its relation to the physical world.

This memoir is of considerable importance. Whitehead thought it one of the best pieces of work he had done.[3] His geometrical and physical interests are combined there with his studies in mathematical logic, which were to see their fruition in *Principia Mathematica*. In it are also to be found the germs of both his nature philosophy and his metaphysics. The axiomatic method is used here in all its rigour. Despite the importance

---

[1] Read before the Royal Society, December 7, 1905. Published in the *Philosophical Transactions of the Royal Society*, 1906. All references will be to the reprint in *Alfred North Whitehead: An Anthology*, edited by F.S.C. Northrop and M.W. Gross, Cambridge, 1953, pp. 11-82.

[2] Russell in *My Philosophical Development*, (Allen & Unwin, 1969) after discussing the view that matter was a plenum pervading all space, states: "This book (Clerk Maxwell's) had been the subject of Whitehead's Fellowship dissertation", and Whitehead urged Russell to prefer its views to those of Boscovitch. When Russell adopted this view he gave it a Hegelian dress "and represented it as a dialectical transition from Leibniz to Spinoza, thus permitting myself to allow that I considered the logical order to prevail over that of chronology" (p. 43).

[3] Cf. *The Philosophy of Alfred North Whitehead*, ed P.A. Schilpp. Victor Lowe, "Whitehead's Philosophical Development," cf. p. 34.

of this memoir it made little impact on the world of learning at the time of publication. This is not suprising: expressed in *PM* symbolism it must have seemed somewhat esoteric when it appeared four years before *PM* was launched upon the world. What must be remembered is that this was not the product of the idle fancy of a young man. Whitehead was then forty-five and had twenty years' teaching and research behind him.

He tells us that Peano's chief symbols are used in the memoir. As he was still working on *PM*, he went on to say: "The changes and developments from Peano, which will be found here, are due to Russell and myself working in collaboration for another purpose. It would be impossible to disentangle our various contributions."[4]

Not very many philosophers have indicated the connection of *MC* with Whitehead's later work. It also seems to have had little effect on physical thought. The axiomatisation of physics, apart from attempts by von Neumann and Reichenbach in the field of quantum theory, has not proved to be acceptable by physicists, possibly because of the rapid advances made in their subject. However, Whitehead is not trying to formalise a fragment of some specific physical theory such as quantum mechanics, but the foundations of the subject needed for any physical theory. He is not blind to the fact that the working physicist relying largely on insight and intuition, may not be willing to accept such a formal basis for his science. In physical research, he points out, so much depends on a trained imaginative intuition that it seems unlikely that existing physicists would gain by deserting familiar habits of thought.[5]

At the time of writing *MC*, Whitehead was not only working on *PM*, he was also about to publish his texts on Descriptive and Projective Geometry. *A Treatise on Universal Algebra* in which he tried to show the connection between logic and geometry, had already appeared some eight years earlier.[6] But it was to be another ten years before his paper "Le Théorie Relationniste de l'Espace,"[7] in which the *Method of Extensive Abstraction* is stated for the first time, was to be seen in print.

This memoir has had little effect on the interpreters of Whitehead's later

---

[4] *MC*, p. 18.

[5] *MC*, cf. p. 12. On another score Whitehead's memoir must have seemed very academic at the time of writing – namely, in the application of logical systems to concrete subject matters. Logical systems have now been applied to a large range of topics; genetics, nerve-networks, computers, etc. Whitehead was a pioneer in this field.

[6] *A Treatise on Universal Algebra*, p. 32. "The result of it is that a Treatise on Universal Algebra is also to some extent a treatise on certain generalised ideas of space."

[7] *Revue de Métaphysique et de Morale*, Vol. 23, May 1916, pp. 423-454.

philosophy, although in some ways it is much more of a speculative endeavour than is his philosophy of nature. He was constructing what in effect
were different cosmologies – a topic he comes back to in his later writings.
MC is an early precursor of *Process and Reality* which is significantly sub-
titled "An Essay in Cosmology."

II. THE AXIOMATIC METHOD

Whitehead in *MC* might be said to be engaging upon a mathematical
(or logical) investigation of various possible ways of conceiving the nature
of the material world. Different axiom systems, or models of the material
world, are constructed, from each of which geometry is derivative. The
axioms from which in each concept of the material world geometry is
derived, are themselves defined in purely logical terms. There seems a certain
resemblance to Leibniz's doctrine of many possible worlds, each of which is
self-consistent, except that in *MC* we are limited to five main types and a
number of variants.

Whitehead makes it clear that he is engaged in constructing logical
models, rather than describing some actual physical situation. He points
out that he is not concerned with (a) the relation of the concept of the
material world to a perceiving mind, or (b) the philosophical problem of the
relation, of any, or all, of these concepts to existence. His account is a purely
hypothetical one.[8]

In writing this memoir, Whitehead seems to have been influenced by
the concept of alternative systems of geometry, in which, starting from
different postulates, we can build up various non-Euclidean geometries.
It was once assumed that Euclid's geometry was based on self-evident axioms
from which all the theorems followed deductively. The elaboration of non-
Euclidean geometries in which the parallels postulate does not hold, has
shown that different geometrical systems can be constructed upon axioms
other than Euclid's. In a somewhat similar manner Whitehead constructs a
number of alternative cosmologies which are hypothetical in character, and
from each of which Euclidean geometry may be constructed.[9]

---

[8] *MC*, cf. p. 13.

[9] Non-Euclidean geometries are also to be constructed if somewhat different axioms are
used.

III. THE DEFINITION OF A CONCEPT OF THE MATERIAL WORLD

Whitehead explains what he means by a "Concept of the Material World":[10] it is in a sense a two-fold notion. Each such concept can be expressed in terms of (a) a set of fundamental relations and the axioms they satisfy, and (b) the class of entities which form the fields of these relations, i.e., the class of ultimate existents – points, lines, instants, etc.

The fundamental relations belonging to each concept are: (1) *The Essential Relation*, (2) *The Time Relation* and (3) *The Extraneous Relation*. (1) is a single polyadic relation of linear order by means of which the various geometrical entities are defined. (2) is a dyadic serial relation similar to that generating the series of positive and negative numbers, and having for its field the instants of time. (3) deals with the way material particles are related to points and instants of time, and in terms of which the laws of dynamics are determined.[11]

Whitehead, as we have noted, constructs five main concepts of the material world. He starts with the classical Newtonian concept and ends with specially constructed concepts where the basic elements are linear entities rather than points of space. In each concept he is primarily concerned with the essential relation and the axioms defining it, which take on different properties according to the particular concept involved. Whitehead states this problem abstractly as follows: "Given a set of entities which form a field of a certain polyadic (i.e., many-termed) relation R, what 'axioms' satisfied by R have as their consequence, that the theorems of Euclidean geometry are the expression of certain properties of the field of R?"[12] He is trying to show how from a certain set of entities and the axioms concerning the relations between them, geometry follows as expressing certain properties of these entities.

It will be noted that the essential relation appears in one form or another in every concept of the material world. There is some resemblance here to the extensive concepts of *PR*, which he thinks are persistent characteristics to be found in every cosmic epoch. These are exemplified in the notion of an

---

[10] *MC*, cf. p. 13.

[11] The essential and extraneous relations would seem to be forerunners of the essential and contingent relations of his later philosophy. In his later work the essential relations relate events to other events in nature, whilst the extraneous relations concern the objects implicated in these events.

[12] Though he adds (*MC*, p. 11), "In view of the existence of change in the material world, the investigation has to be so conducted as to introduce, in its abstract form, the idea of time, and to provide for the definition of velocity and acceleration."

Extensive Continuum, which we are told, "is one relational complex in which all potential objectifications find their niche."[13]

Whitehead now outlines the logical procedure to be followed in the study of any concept of the material world, and for that matter, he adds, any investigation respecting the axioms of geometry or physics viewed as deductive sciences. Four stages are distinguished. These are: (1) The definition of the entities capable of being defined in terms of the fundamental relations; e.g., a straight line is defined in terms of a class of points between which the essential relation R holds. (2) The deduction of the various properties implicit in those definitions, e.g., statements about lines and planes, derived from the above. (3) The selection of the group of axioms which determine that concept of the material world; e.g., axioms concerning the properties of R. (4) The deduction of propositions from the initial axioms, e.g., the ordinary propositions of geometry which in Euclid are the derived theorems.

Psychologically, he points out, the order of these studies tends to be inverted. We first choose propositions of stage (2) and (4) and then go on to stages (1) and (3). What Whitehead is essentially concerned with in each such concept is to exhibit stages (1) and (3), i.e., the selection of the definitions and axioms of the system. Or putting it in another way, in our ordinary experience we are intuitively familiar with the properties of geometrical figures. When, however, we put geometry (or any other concept) on to a purely formal basis, the geometrical propositions are now shown to be derivable from more basic propositions. They can, as, for example, in Veblen's axiomatisation of geometry, be constructed from a set of points and a triadic serial relation.

IV. DUALISTIC AND MONISTIC CONCEPTS

Whitehead first gives a critical examination of the classical Newtonian concept of the material world. This, he says, arose in an age when geometry as derived from the Greeks was the only developed science. The application of geometry to the physical world led to the acceptance of the absolute theory of space, upon which was superposed a theory of the motion of matter in time.

Three mutually exclusive classes of "objective reals" are then assumed

[13] On the definition of the Extensive Continuum see *PR*, p. 91. The relation of extensive connection of *PR* seems to have developed from the simple notion of intersection which is a property of the essential relation of Concept V.

by this concept: points of space, instants of time and particles of matter.[14] To these correspond three sciences. geometry, which concerns itself with the relations between points; chronology – the theory of time as a one-dimensional series ordinally similar to the series of real numbers; and dynamics, which deals with the movement of matter in space and time.

Whitehead objects to the classical concept on the ground that it bifurcates the world into separate elements, namely, space and the matter occupying it. It thus involves an unnecessary duplication of physical entities. Starting from Occam's principle that "Entities are not to be multiplied without necessity,"[15] Whitehead believes that there is an intuitive preference for a monistic concept of the material world. In a monistic concept space and matter are unified so that only one class of entity can occur. Such a concept is termed by him Leibnizian, since for Leibniz points of space are not conceived as existing in themselves, but rather as denoting an order of co-existing things.

A somewhat similar criticism of the classical concept is found in *An Enquiry Concerning the Principles of Natural Knowledge* (*PNK*), where by way of an introduction to his own philosophy of nature, Whitehead compares the classical concept with what he calls the "principle of relativity." He links this up with his question, "What is a physical explanation"? The orthodox reply, he answers, was in terms of "Time (flowing equably in measurable lapses) and of Space (timeless, void of activity, euclidean), and of Material in space (such as matter, ether or electricity)."[16] Since the same material exists at different times, the notion of a state of change must be included.

On the other hand, according to the "principle of relativity" which forms the basis of his own view, the properties of space are merely a way of expressing relations between things ordinarily said to be in space. All spatial entities, such as points, straight lines and planes are mere complexes of relations

---

[14] Whitehead takes over Russell's logical analysis of the nature of matter. Cf. *The Principles of Mathematics*, p. 468. On this analysis (1) a simple material unit occupies a spatial point at the same moment and persists through time, (2) two units cannot occupy the same point at the same moment, (3) their positions at intermediate times must form an intermediate series. Matter itself is a collective name for all pieces of matter, as space is for all points and time for all instants.

[15] See *The Aims of Education*, p. 218, where we are told that the scientific validity of Occam's razor *Entia non sunt multiplicanda praeter necessitatem*, is obvious, "namely, every use of hypothetical entities diminishes the claim of scientific reasoning to be the necessary outcome of a harmony between thought and sense-presentation. As hypothesis increases necessity diminishes."

[16] *PNK*, p. 1.

between things. Starting from this position, spatial concepts are defined in terms of the relations between material things.

Whitehead in *MC* also contrasts the classical concept with Leibniz's theory of the relativity of space. From a wider standpoint, however, Whitehead sees this theory as a protest against dividing the "objective reals" into two parts: on the one hand, points of space unrelated to instants of time and on the other, material particles which have a time-reference. In this sense, he states, it may be said to be a protest against exempting any part of the universe from change.[17] Not only is he then criticising the view that space can exist apart from things, but also that it can exist unchanging and out of time.

In effect, what this comes to on a monistic concept of the material world, is that instants of time are included as terms within the essential relations from which geometry is derived. In such a system geometrical statements have therefore a time reference. There is already a certain similarity here to Whitehead's later doctrines that geometrical forms are derivative from process, and that the fundamental entities are events which change and pass. But there is this difference: space and time are still regarded as independent of each other. They are not as in his nature philosophy fused together in the form of events.

V. PUNCTUAL CONCEPTS

The first concept of the material world which Whitehead discusses, Concept I, is a formalisation of classical Newtonian physics. It is a dualistic concept since it posits both points of space and material particles having a time reference. Whitehead shows how Euclidean geometry may be derived from this model by taking over Veblen's axiomatic system, which bases Euclidean geometry on one class of undefined entities – points – and an undefined serial relation between points, and adapting it to his concept of an essential relation. In Concept I the essential relation R is triadic and is symbolised as "R; (*a b c*)" – i.e., *the points a, b, c are in the linear order (or the R-order) abc*.[18] The relation, we are told, may be varied so as to make the resultant geometry Euclidean or non-Euclidean.

But geometry thus derived has to apply to a changing world. Whitehead therefore postulates a class of particles which make up the fundamental stuff moving in space, and a class of triadic extraneous relations holding

---

[17] *MC*, cf. p. 14.

[18] *MC*, cf. p. 25. Twelve axioms concerning this relation are stated, and from which he is enabled to derive Euclidean geometry.

between a particle, a point of space and an instant of time. These extraneous relations are superposed upon the geometrical entities and provide for the definition of velocity and acceleration in terms of which the laws of dynamics are defined.

Concepts II and III as concepts of the material world are monistic variations of Concept I. In both concepts geometry is brought into closer relationship with physics. In Concept II a material particle becomes a mode of relating a point of space to an instant of time. Russell, who suggested this concept, put it as follows: "It is plain that the only relevant function of a material point is to establish a correlation between all moments of time and some points of space, and that this correlation is many-one."[19] As Whitehead believes that "matter" was only introduced to give the senses something to perceive, this concept, which reduces "matter" to the form of a relation, has for him an advantage over the classical one, if relations can be perceived.

In Concept III (of which two variants are stated) the essential relation from which geometry is derived is a tetradic one. It is symbolised as "R; (a b c t)" – i.e., *the objective reals a, b, c are in the R-order at the instant t*,[20] where the objective reals are the points of the classical concept. The geometrical definitions occurring in this concept are the same as in Concept I except that any geometrical proposition now involves a reference to an instant of time. A distinctive feature of Concept III is that the points of this concept are assumed to be in motion. The whole of space is made up of moving points (or particles), the positions of which change at different instants. The persistence of matter as exhibited, for example, in the endurance of an electron, is hence solely to be explained by the persistence of motion.[21] Whitehead also introduces a single extraneous relation S which enables him to define velocity and acceleration.

VI. LINEAR CONCEPTS

The last two concepts of the material world discussed by Whitehead are Concepts IV and V. In these concepts the terms related by R (the essential relation) are linear in character and points are classes of these simple linear entities. He conceives a linear objective real as a single unity underlying a

---

[19] *The Principles of Mathematics*, p. 468.

[20] *MC*, p. 30.

[21] *MC*, p. 31. This suggests his later view of a relatively permanent body as a succession of events having a similar pattern.

straight line which has a particular direction (i.e. a vector character). White-head empirically identifies these linear entities with "lines of force" in physics, which he takes to be "the ultimate unanalysable entities which compose the material universe,"[22] and "geometry is to be regarded as a certain limited set of these properties."[23] He further assumes that the totality of linear reals taken as a system, will like the lines of force making up the material world, form a field or continuum.

The essential relation from which geometry is derived in Concepts IV and V is a pentadic one, and is symbolised by "R; $(a\ b\ c\ d\ t)$" – i.e., *the objective real a intersects the objective reals b, c, d in the R-order at the instant t*.[24] The notion of intersection introduced here has some resemblance to his later notion of extension in his philosophy of nature (i.e., the way each event overlaps – or extends over – other events in nature). In Concepts IV and V geometry has then not only to do with points of space, but also with their relation to the linear reals from which they are derived. If we were to inject time or passage into these linear entities, we would come pretty close to Whitehead's later doctrine of events. It is interesting to note that in *PR* the fundamental experiential elements or "prehensions" also have a vector character. Two variants of Concept IV are stated by Whitehead.

In his account of the linear concepts Whitehead postulates a class of entities which he believes corresponds more closely to the basic physical elements than do points. At the time of writing *MC* he thought that Max-well's electromagnetic theory had made a linear concept more likely than a punctual one. But in such a later work as *AI*, he realises that in more recent physics atomicity in the form of electrons and protons has again emerged into importance. Nevertheless, he believes that these two contrasted aspects of nature, continuity and atomicity, are both essential for our descriptions of nature.[25]

VII. CONCEPT V

In Concept V, which is a monistic one, Whitehead derives geometry from postulated linear reals, which can be given an empirical interpretation in terms of physical notions (i.e. lines of force). As in Concept IV the essential

---

[22] *MC*, p. 32.

[23] Compare the above with his statement (*AI*, p. 258), "The peculiar relationships (if any) diffused systematically between the extensive groups of an epoch constitute the system of geometry prevalent in that epoch."

[24] *MC*, cf. p. 39.

[25] *AI*, cf. p. 239.

relation from which geometry is derived is pentadic: one extraneous relation is also required for the purpose of allowing velocity and acceleration to be measured. Seventeen axioms are postulated in contrast to the twelve of Concept I, since the field now to be covered is larger and includes physics. Thus in a monistic concept, although the number of undefined entities is reduced there may be an increase in the number of axioms and derived theorems.

Whitehead in *MC* and particularly in Concept V is concerned with the question: "How can a point be defined in terms of lines?," since he wishes to derive points from linear reals which he thought were closer than points to the ultimate physical existents. He introduces two methods of doing this. The first is *The Theory of Interpoints* which seems to be a precursor of his *Method of Extensive Abstraction.* In this theory a point is defined in terms of a class of entities having a similar property – namely, that of occupying the same position.[26] In a similar fashion Leibniz, faced with the problem of defining the notion of place on a relational theory of space, defined place in terms of a series of entities each of which could occupy the same place. Another way of defining points is also stated in *The Theory of Dimensions* where Whitehead constructs a three-dimensional geometry applicable to the physical world. Starting from our ordinary geometrical ideas, he proceeds to define points in the way an ideal point is defined in projective geometry. A point in this theory becomes simply "the class of lines concurrent at a point."[27]

At the end of *MC* Whitehead sketches a possible development of Concept V. "The complete concept," he tells us, "involves the assumption of only one class of entities as forming the universe. Properties of 'space' and of the physical phenomena 'in space' become simply the properties of this single class of entities." In order to simplify the axioms of Concept V, "the ideal to be aimed at would be to deduce some or all of them from more general axioms which would also embrace the laws of physics. Thus these laws should not presuppose geometry, but create it."[28]

We seem to be given here a preview of his later position that geometrical relations arise from the physical details of the cosmic epoch in which we live. In *PR* this shows itself in his belief that the order of nature presupposes the mutual interdependence of the real things to be found in nature. Since the laws of nature depend on the characters of these things, as the things change so will these laws also change.

[26] *MC*, cf. p. 35.
[27] *MC*, p. 48.
[28] *MC*, p. 82.

What should be noted is that although in Concept V Whitehead says he is putting forward a monistic concept, it is only monistic in the sense that it does not postulate two types of entities, points and material particles. However, this concept is dualistic in another sort of way, since it postulates both essential and extraneous relations, as does the classical concept. There is thus still a dualism between the relations determining geometrical order and the relations determining changes of motion in the world, and from which the laws of dynamics are derived. It is interesting to note that a similar dualism is to be found in his later writings when he compares the world of permanences and the world of change.

In view of the statistical basis of modern physics and the substitution of chance for cause in physical explanations, it is doubtful whether physicists would accept such a frankly causal account as that of Concept V nowadays. Concept V and the whole tenor of *MC* is determined by the fact that Whitehead conceives nature as continuous and made up of logical linear entities resembling lines of force. Whitehead's model is constructed with Clerk Maxwell's theory of electromagnetism in mind, which seems to fit more naturally on to a causal macroscopic level than a statistical microscopic level.

The lesson to be learnt from this is that as physics advances so one's models require to be adjusted to changes in physical theory and subject matter. Whitehead recognises this in *Adventures of Ideas* (*AI*) when he admits that the more punctual discontinuous side of nature has shown itself again in quantum physics. He believes, however, that his theory can take account of such discontinuities in nature. And in *PR* he states that his cosmology is consistent with the view that physical laws are statistical in character.

Whitehead would also argue that the basic elements assumed in physics – space, time and matter – have remained the same, though the kinds of matter postulated and the laws they follow may have changed. It is these basic elements that he is trying to express in logical terms. Whitehead's argument is then a logical one: to reduce the number of basic concepts physics starts from. He might argue that the advances in physics have led to the need of a change in the different kinds of extraneous relations in terms of which the physical laws are stated, rather than in the essential ones from which geometry is derived. If this is the case, the *MC* account suitably modified should be acceptable to different physical theories of nature.

III. THE RELEVANCE OF THE MEMOIR TO WHITEHEAD'S LATER PHILOSOPHY

Although it has been stated that this memoir has a relevance for White-head's later philosophy, precisely what this relevance is has not always been made clear. Quine, for example, tells us, that Whitehead was occupied in analysing the logical structure of natural science, and that the constructions foreshadow to some degree the projected *PM* volume on geometry. As we have seen, it is largely an account of different models of the physical world (i.e. of alternative possible worlds) and of how in each case Euclidean geometry may be constructed by means of purely logical notions. Quine notes this when he points out that the constructions go far outside geometry: "this was the beginning of a quest for the broadest, most basic concepts and principles of nature, and in the decades since *Principia*, the quest has issued in a metaphysics."[29]

One thing is clear: if *MC* had been discussed in any detail at an earlier date, it might have been noted when *PR* appeared, that there was some connection between the method of speculative philosophy in *PR*, and the axiomatic method used to construct the various concepts of the material world in *MC*. Those commentators who recognise that there is a link between his nature philosophy and *PR* have tended to start with his nature philosophy writings, where we seem to be given a form of phenomenalism, and where the notion of alternative systems of the world is not very evident. It might indeed be argued that Whitehead's philosophy of nature with its emphasis on the "principle of relativity," is largely a development of Concept V of *MC*, to which an experiential dimension has been added.

Students of Whitehead's later metaphysical writings have pointed out that there has been a progression in these writings beyond his down-to-earth nature philosophy. They have argued quite rightly that *PR* stresses something which seems absent from his earlier nature philosophy. In his earlier work the part played by imaginative generalisation and speculation in philosophical thinking, does not seem to have been much discussed. No doubt it was this down-to-earthness which gave Whitehead's nature philosophy its appeal to the more positivistically minded philosophers, and also explains their consternation when *PR* appeared.

In *PR* the proper objective of philosophy is conceived as the gradual elaboration of categoreal schemes, definitely stated at each stage of pro-

---

[29] *The Philosophy of Alfred North Whitehead*, ed. P.A. Schilpp, Willard V. Quine, "Whitehead and the Rise of Modern Logic," p. 163.

gress. Metaphysical categories, we are told, "are tentative formulations of the ultimate generalities." When we have constructed such a scheme, we need to put it into use by arguing from it boldly and with rigid logic. For this reason the "scheme should therefore be stated with the utmost precision and definiteness, to allow of such argumentation."[30]

There would seem to be a family resemblance between his account of philosophical method in *PR* and the method used for the construction of a concept of the material world in *MC*. In *MC* Whitehead tells us, that he discusses the general problem involved in the construction of concepts of the material world, purely for its logical (i.e. mathematical) interest. It has an indirect bearing on philosophy "by disentangling the essentials of the idea of a material world from the accidents of one particular concept."[31]

In his Preface to *Science and the Modern World* (*SMW*) Whitehead has said that "Philosophy, in one of its functions, is the critic of cosmologies."[32] Now although only one cosmology is developed in *PR*, Whitehead does discuss alternative cosmologies. He compares the cosmology of Newton as it appears in the *Scholium* to the first eight definitions of his *Principia*, with that of Plato as expressed in the *Timaeus*.[33] The former is, of course, the classical concept of *MC*, whilst the latter is a monistic one resembling Concept V.

Whitehead points out that in the *Scholium* space and time are ready-made for the material masses, which are themselves ready-made for the "forces" which constitute their action and reaction.[34] But there is, he tells us, a side in the *Timaeus* which finds no analogy in the *Scholium*, which may be termed its metaphysical character, "its endeavour to connect the *behaviour* of things with the *formal nature* of things."[35] Or, as he states elsewhere in *PR*, geometry is derivative from the physical societies constituting any epoch. The *Timaeus* type of cosmology resembles that of Concept V, where geometry is derivative from quasi-physical notions and where the behaviour of things is connected with the formal nature of things.

Although Whitehead's metaphysics undoubtedly has a certain resemblance to the position of *MC*, from which his nature writings developed, it nevertheless, has a definite experiential character. In his later philosophy

[30] *PR*, p. 11.
[31] *MC*, pp. 11-12.
[32] *SMW*, p. 7.
[33] *PR*, cf. p. 129.
[34] *PR*, p. 130.
[35] *PR*, p. 131.

we deal, among other things, with questions relating to sense-perception, consciousness, judgment, etc., which are not touched upon in *MC*.[36]

---

[36] Even in *PR* (cf. pp. 126-7) Whitehead conceives cosmology as having a somewhat arbitrary character. In this epoch, he tells us, "Maxwell's equations of the electromagnetic field hold sway .... But, the arbitrary factors in the order of nature are not confined to the electromagnetic laws. There are the four dimensions of the spatio-temporal continuum, the geometrical axioms, even the mere dimensional character of the continuum ... it will be evident that all these properties are additional to the more basic fact of extensiveness; also, that even extensiveness allows of grades of specialisation, arbitrarily one way or another."

CHAPTER 5

THE PHILOSOPHY OF NATURE

I. NATURE AND PERCEPTION

Whitehead's philosophy of nature is largely presented in his books *An Enquiry concerning Natural Knowledge* (*PNK*) and *The Concept of Nature* (*CN*), and to some extent in *The Principle of Relativity* (*P of R*), although the latter is also concerned with his theory of relativity. One of the reasons for the neglect of these works in recent years is that philosophers of science have tended to be interested in other types of problems than those discussed by Whitehead there. They have been primarily taken up with questions relating to the linguistic analysis of science and with models of scientific theories. It is a little ironical to reflect that this approach is probably due to the influence of mathematical logic, of which Whitehead was one of the pioneers. In his philosophy of nature Whitehead, however, was concerned with a very different sort of problem, which he would have considered more fundamental, namely, the relating of the abstract notions of physics to our sense-experience.

Whitehead points out[1] that modern developments in theoretical physics and particularly relativity theory have made it necessary for us to look into the question "What are the ultimate data of science?" It is, he says, of the nature of things that we first act and then proceed to discuss the *rationale* of our activities. Thus the creation of science precedes the analysis of its data.[2] Although science can continue with faulty analyses, Whitehead believes that the acceptance of such analyses can only lead to a warping of the scientific imagination. The starting point of his enquiry, he tells us, is Nature as disclosed to us in perception. He therefore considers the primary tasks of a philosophy of science to be (1) to exhibit the fundamental entities and fundamental relations between them which we discern in nature, and (2) to secure

---

[1] *PNK*, p. iv.
[2] *PNK*, p. iv.

that the entities and relations thus exhibited are adequate for the expression of all the entities and relations which occur in nature.

Traditional scientific theory, on the other hand, he tells us, interprets natural phenomena as a function of material bodies situated in space and time. This approach to nature, he argues, shows the influence of Greek philosophy on science,[3] particularly Aristotelian logic[4] where the fundamental type of affirmative proposition is the attribution of a predicate to a subject. This approach has given rise to the view that nature consists of substances manifesting attributes or qualities, and has led to the engrained tendency to postulate a substratum for whatever is disclosed in sense-awareness.[5] The result has been that the entity deprived of all characteristics except those of space and time, has acquired a physical status, and the course of events has been conceived as being merely the fortunes of matter in its adventures through space.[6]

Nevertheless, Whitehead believes that the doctrine of matter does enshrine some fundamental law of nature, although the scientific expression of this fact has become entangled in a maze of doubtful metaphysics. When this metaphysics is put aside, he says, a new light is thrown on many fundamental concepts which dominate science. It will then be seen that our usual way of approaching natural phenomena is in terms of high abstractions. The acceptance of such abstractions as ultimate realities has led to what he terms the bifurcation of nature. Nature has been bifurcated into two systems of reality.[7] There is (1) apparent nature which holds "within it the greenness of the trees, the song of the birds, the warmth of the sun, the hardness of the chairs and the feel of the velvet."[8] And (2) causal nature "the conjectured system of molecules and electrons which so affects the mind as to produce the awareness of apparent nature."[9] But, Whitehead argues, in considering knowledge we need to wipe out such spatial metaphors as "within" or "without" the mind. We cannot, he says, explain the "why" of knowledge: we can only describe the "what" of knowledge, the entities and relations we find in nature as given in sense-experience.[10] We cannot explain nature, as it is the ultimate datum from which we start.

---

[3] *CN*, p. 16.
[4] *CN*, p. 18.
[5] *CN*, p. 18.
[6] *CN*, p. 20.
[7] *CN*, p. 30.
[8] *CN*, p. 31.
[9] *CN*, p. 31.
[10] *CN*, p. 33.

On the materialist theory, Whitehead points out, it becomes especially difficult to bring together in one system the perceived redness and warmth of the fire and the agitated molecules of carbon and oxygen and the various functionings of the material body which are said to be causal of these sense-qualities.[11] The distinction drawn between "apparent" and "causal" nature means that they are conceived as belonging to radically different kinds of reality. It is therefore not suprising that philosophical attempts to bring about their union have proved unsuccessful. The materialist approach to reality also encounters difficulties when it is applied to biological organisms, as an organism cannot be expressed in terms of a material distribution at an instant.[12] "The essence of an organism," Whitehead tells us, "is that it is one thing which functions in and is spread through space."[13] As thus conceived a biological organism is obviously incompatible with the traditional theory of materialism. And he goes on to argue that material objects like organisms also need to be regarded as unities with a spatio-temporal extension: a piece of iron, for example, requires a period of time for its functioning.

Another assumption implicit in the materialist theory is that it conceives time as a series of self-contained instants, completely independent of the things said to occur in time. On this view Whitehead states "the instantaneous present is the only field for the creative activity of nature ... this unique present is the outcome of the past and the promise of the future."[14] But Whitehead denies that we are immediately aware of such an instantaneous present. What we experience is what he terms a duration – a present whole of nature – which contains within itself a past and a future. A duration also implies its apprehension by a percipient event, which is, he says, roughly "the bodily life of the incarnate mind." The special relation of the percipient event which is "here" to the duration which is "now" is termed by him cogredience,[15] and may be said to describe the percipient event's peculiar point of view within the whole of nature given in sense-awareness.

Whitehead tells us that the basic elements discriminated by us in nature are events. Even those entities which do not seem to exhibit passage are events. If, for example, we consider Cleopatra's Needle on the Thames Embankment, it seems at first sight to lack the transitoriness of an event. But the static timeless element in the relation of Cleopatra's Needle to the Em-

---

[11] *CN*, p. 32-33.
[12] *PNK*, p. 3.
[13] *PNK*, p. 3.
[14] *CN*, p. 72.
[15] *PNK*, p. 71.

bankment is pure illusion. What it really comes to, he says, is that amidst the structure within which the daily life of Londoners is passed, we know how to identify a certain stream of events which maintains a permanence of character, that of being the situation of Cleopatra's Needle. If the Needle is defined in a sufficiently abstract way, we can say that it never changes. But a physicist will tell us that it has lost some molecules and gained others, and even the plain man can see that it gets dirtier and is occasionally washed.[16]

The characters of events which we recognise and which play an important part in making us believe in the permanence of events are termed by Whitehead objects: they are factors which do not participate in the passage of events. He points out that the way in which events and objects enter into experience differ: events are lived through and objects are experienced through the intellectuality of recognition.[17] We recognise the self-same object, although it may appear in different events.[18] It is therefore objects which give events their specific characteristics, and which mark them off from different events. However, the relationship which an object has to an event is not that of an attribute to a substance: it is a multi-termed relationship which he calls *ingression*;[19] and this involves as terms not only the event in which the object appears, but other events in nature including the percipient event.

Whitehead notes that the relation of *ingression* is not necessarily identical with that of situation. Consider, he says, a flannel coat of Cambridge blue belonging to an athlete. Whitehead is not referring here to someone's definite sense-awareness of Cambridge blue but to the situation of the actual event itself. If one is looking at the event directly, one will then see Cambridge blue as situated practically in the same event as the coat itself at that instant. It is true, Whitehead goes on, that the blue which one sees is due to the light which left the coat some inconceivably small fraction of a second before, a difference which would be important if one was looking at a star whose colour is Cambridge blue, since the star might have ceased to exist millions of years ago.[20]

Our perception of a Cambridge blue coat as situated in a certain event (or as ingressing in it), he goes on, therefore involves a multiple relationship. The *relata* in this relationship are (a) a percipient event (the relevant bodily states of the observer); (b) the situation in which the objects are

[16] *CN*, pp. 166-167.
[17] *CN*, p. 143.
[18] *PNK*, p. 64, *CN*, p. 144.
[19] *CN*, pp. 144-148, *PNK*, pp. 83-88.
[20] *CN*, p. 151.

situated (where one sees the blue); (c) the active conditioning events (in this case, the coat, the state of the room as to lighting and atmosphere); and (d) the rest of the events of nature, which he calls the passive conditioning events.[21] In general, he tells us, the situation is an active conditioning event, namely, the coat itself, when there is no mirror, etc., to produce abnormal effects.[22] In the latter case, there will be a disconnection between the situation of the perceived blue (i.e. as behind the mirror) and the situation of the actual object. Hence, the relation of situation (or ingression) and that of causal influencing are not necessarily the same sort of relation. Whitehead believes that the difficulties which cluster round the relation of situation arise from the influence of Aristotelian logic, which attempts to reduce all relationships to the substance-attribute relation.[23] Because of this, he says, philosophers refuse to take seriously the ultimate fact of multiple relations.

Whitehead proceeds to classify the most important types of objects of which we have knowledge. These are: (1) sense-objects, which are the simple self-identical permanences which we find in perceived nature; (2) perceptual objects, which are associations of sense-objects in the same situation: these are the objects of our daily experience, trees, chairs, tables, etc. We have an immediate awareness of this type of object, which is not primarily the result of a judgment. Then (3) there are scientific objects, i.e., molecules, electrons, etc., and these embody those characteristics of situations (where physical objects are usually said to be located) which are most permanent and are expressible without reference to a multiple relation involving a percipient event. They are of a higher degree of abstraction than the other two kinds, which they presuppose.[24]

In order to take account of illusory perceptions Whitehead distinguishes two kinds of perceptual objects: delusive perceptual objects and physical objects – the ordinary objects which we perceive when our senses are not cheated. The situation of a delusive object, for example, the blue coat seen at the back of the mirror, is, he tells us, a passive condition for its appearance: it will also only be the situation of that object for one particular percipient event. Further, when we see the blue image in the mirror, we as it were, complete it by means of what Whitehead calls "conveyed" or subconscious sense objects.[25] This process also occurs in normal perception. We may feel ourselves as wearing or touching the coat, or when we see the

[21] CN. p. 152.
[22] CN, p. 153.
[23] CN, p. 150.
[24] CN, pp. 148-149, PNK, Chap. VII, pp. 82-89.
[25] CN, pp. 154-155.

front part of a box, we may also be conscious that it has a back to it. When we therefore see the blue mirror image, we make the same subconscious response to it as we would to the actual coat itself, and it is because of this that we say we have been deceived.

Whitehead points out that events as discriminated by us in perception have to each other a certain general relation which he terms extension. It is a part and whole relation and is hence transitive. Thus if x is an event and x extends over y and y over z, then y and z are both contained in x. Extension, Whitehead tells us, is a relation so simple that we hardly recognise it as such. Thus "the event which is the continued existence of the house extends over the event which is the continued existence of a brick of the house, and the existence of the house during one day extends over its existence during one specified second of that day." Every event then extends over (or can contain) other events,[26] and it is the relation of extension which gives events their continuity. It is, he says, by means of the properties which issue in spatial relations that the relation of extension exhibits events as actual, as matters of fact, and by means of its properties which issue in temporal relations that it exhibits events as involved in the becomingness of nature – in the passage, the creative advance of nature.

One of the primary aims of Whitehead's enquiry is to show how the scientific concepts of space and time are rooted in sense-experience. They are not, he says, to be "looked for at the tail end of a welter of differential equations."[27] In classical Newtonian physics, geometrical elements such as points and lines as well as instants of time have an independent existence in nature. But as Whitehead points out, relativity theory has made the belief in this view implausible. He goes on to say that he is concerned here with physical space rather than with the various ideal spaces investigated by the mathematician: "We now know many alternative sets of axioms from which geometry can be deduced by the strictest deductive reasoning. But these investigations concern geometry as an abstract science deduced from hypothetical premisses. In this enquiry we are concerned with geometry as a physical science."[28] Such an enquiry has "to explain space as a complex of relations between things. It has to describe what a point is, and has to show how geometric relations between points issue from the ultimate relations between the ultimate things which are the immediate objects of knowledge."[29] And to do this it has to start with a discussion of the character of the immediate data of perception.

[26] *PNK*, p. 75.
[27] *PNK*, p. vi.
[28] *PNK*, p. v.
[29] *PNK*, p. 5.

II. THE METHOD OF EXTENSIVE ABSTRACTION

As we have seen Whitehead pictures nature as an interrelated structure of events, each event possessing volume and duration and having to each other the part and whole relation of extension which is more general than the specific notions of space and time. The key notion used by the *Method of Extensive Abstraction*, which he hopes will enable us to arrive at ideal points and instants, is that of convergence with diminution of extent. Starting with some large enough event we can analyse it into a convergent series of successively smaller events rather like the children's toy consisting of boxes each fitting within the other; except that unlike the toy there is in the case of events no terminal event – only a fictional limit.[30] As the abstractive set converges there is a progressive diminution in the extent – spatial and temporal – of the events considered, so that we finally arrive at the ideal of an event, so restricted as to be without extension in space and time. Whitehead calls such an ideal restricted event an event particle – a mere "point-flash" of a duration. We can express the spatio-temporal structure of physical nature in terms of such event particles, or the points of instantaneous space.

The method has however, been criticised on a number of grounds. Among the objections which have been raised against it are the following: (1) geometrical elements, namely, points, lines, volumes and areas, are presupposed in the definitions of an abstractive set; (2) we never perceive the smaller events of the convergent series; and (3) the method presupposes some sort of conceptual extrapolation to deal with the condition of inclusion, which presupposes the notion of an infinite series.

One of the reasons for the difficulties to which Whitehead's view gives rise, is that he does not make clear whether his method is to be taken as an algorithm or as a description of some actual process of convergence starting from our perceptions. According to Russell, Whitehead was examining from the point of view of mathematical logic how we may deduce in terms of empirical data the entities that traditional geometry considers as primitive. However, this method, Russell goes on, starts from knowledge of the completed mathematical system, which is the object to be obtained, and goes back to entities more analogous to those of sense-perception, i.e., events.[31] Russell made the further comment that in psychological space the method does not yield continuity unless we assume that sense-data which have a minimum size below which nothing is experienced, always contain

---

[30] *CN*, p. 79.
[31] Preface to Jean Nicod's *Foundations of Geometry and Induction*, cf. p. 7.

parts which are not sense-data. Russell concludes that the full employment of Whitehead's methods belongs rather to physical space than to the space of experience.[32]

If Whitehead's method is to be taken as a piece of applied logic, as an abstract model enabling us to get from volumes to points – and which we can use to make clear certain relations appearing in sense-perception, then the question arises how far the regions of the system apply to the sensed volumes: although the regions of the model converge to definite limits, the set of sensed volumes only approximates towards such an ideal simplicity. Whitehead was clearly aware that the abstractive sets could not get us to a terminal event. He did, however, seem to argue that if the sensed volumes had certain properties of the abstract series then we could, as it were, in thought converge the sensed volumes.

Thus Whitehead might defend his position on the ground that we do at least in thought encounter the inexhaustibility of the abstract set. From this it would appear that the one interpretation of the method which seems to agree with Whitehead's general position, is that he is not trying to construct a geometry from sense-experience, but rather using a mathematical model to make clear certain relations appearing in sense-perception. This interpretation would agree with his later account of the method in *PR*, where he starts from the more primitive relation of "extensive connection," which includes that of extension as a special case. But unlike his earlier account, the abstract set from which points are now derived is based on the interconnections of abstract postulated regions.

Russell at one time championed Whitehead's *Method of Extensive Abstraction*. He tells us that he was persuaded by Whitehead that one could do physics without assuming points and instants to be part of the stuff of the world. He went on to say that in this way he thought it possible to show how such structures can be constructed by mathematical logic from the higgledy-piggledy world we live in. "For this reason, mathematical logic is an essential tool in constituting the bridge between sense and science."[33] And as Whitehead himself said of the method, "It is a method which in its sphere achieves the same object as does the differential calculus in the region of numerical calculation, namely it converts a process of approximation into an instrument of exact thought."[34]

---

[32] *Our Knowledge of the External World*, 1949 edition, p. 121.
[33] *My Philosophical Development*, p. 206.
[34] *PNK*, p. 76.

CHAPTER 6

SCIENCE AND THE MODERN WORLD

I. SCIENCE AND PHILOSOPHY

Whitehead's book *Science and the Modern World* (*SMW*),[1] which he published originally in 1925 shortly after he came to Harvard, examines in some detail the development of materialism as a scientific and philosophic doctrine from Greek times to the present day. Whitehead studies the cluster of concepts which make up the materialist doctrine not merely from an epistemological point of view as was the case in *PNK* and *CN*, but he also tries to see them in their historical perspective. He proceeds to show how modern developments in physics and biology – particularly the theory of evolution – have made the materialist doctrine increasingly untenable. And against this background Whitehead sketches an alternative philosophy of organism which he hopes may come to replace materialism.

SMW may then be said to be primarily a study of the growth of science over the past three centuries and its impact on Western culture, with Whitehead being concerned to bring out the important influence science has had on our modes of thought and metaphysical presuppositions. He believes that the scientific attitude is largely based on the instinctive conviction inherited from the Greeks that there is an *Order of Things* and in particular an *Order of Nature*. But since Hume's time, he remarks, it has been fashionable to deny the rationality of nature: on Hume's view of causality science merely establishes entirely arbitrary connections between phenomena.[2]

There would then, Whitehead says, seem to be a basic inconsistency implicit in scientific thinking. Scientists have believed that nature is thoroughly rational – and that it manifests a definite order – although they have been unable to give a logical justification of this belief. He therefore holds that it is of some importance to trace the historical antecedents of the instinctive

---

[1] *Science and the Modern World*, Penguin Edition, 1938, (*SMW*).
[2] *Ibid.*, p. 14.

faith that there is an *Order of Nature*.[3] He notes that the Greeks conceived their cosmology as articulated in the manner of a work of dramatic art, so as to exemplify general ideas converging to a specific end. But the effect of this approach was to damp down the historical spirit.[4] "For it was the end which seemed illuminating, so why bother about the beginning?"[5]

Whitehead argues that the Greek vision of a remorseless and indifferent fate urging a tragic incident to inevitable issue, is the vision possessed by science. "Fate in Greek Tragedy," he says, 'becomes the order of nature in modern thought."[6] However, for science to be possible something more is needed than a general sense of the order in things. Precise analytical thought is also needed, and this science owes to the influence of scholastic logic and scholastic divinity: "the priceless habit of looking for an exact point and sticking to it when found."[7] But what medieval thought particularly contributed to the scientific movement was the inexpungible belief that every detailed occurrence was to be correlated with its antecedent in a perfectly definite manner, so that it exemplified general principles.[8] And this, Whitehead believes, was born from the proper insistence in the Middle Ages on the Rationality of God "conceived as with the personal energy of Jehovah and with the rationality of a Greek philosopher,"[9] so that every detail of the universe was supervised and ordered. This faith in the possibility of science born prior to the development of modern scientific theory is, Whitehead claims, an unconscious derivative from medieval theology.[10]

But if science is to be more than a speculative enterprise it requires not only precise logical thought, but also a reference to the facts themselves. This largely came about during the later Renaissance through what Whitehead calls the historical revolt.[11] This showed itself in a twofold manner: in the field of religion at the Reformation as an appeal to the origins of Christianity, and in science as an appeal to efficient as against final causes. Francis Bacon, for example, argued in favour of induction as giving us new knowledge as against Aristotle's syllogistic, which manifestly did not. But this revolt in science with its emphasis on a return to the facts was, White-

[3] *Ibid.*, p. 14.
[4] *Ibid.*, p. 19.
[5] *Ibid.*, p. 19.
[6] *Ibid.*, p. 21.
[7] *Ibid.*, p. 23.
[8] *Ibid.*, p. 23.
[9] *Ibid.*, p. 24.
[10] *Ibid.*, p. 24.
[11] *Ibid.*, p. 19.

head notes, largely anti-rationalist in character: it was a reaction to the inflexible rationality of medieval thought.

Despite the use of logical and mathematical methods empirical science, Whitehead goes on, has remained predominantly an anti-rationalist movement, based on a naive faith which it has never bothered to justify.[12] Although it claimed to have emancipated itself from philosophy and the restrictive logic of scholasticism, it nevertheless accepted certain philosophical assumptions and these were enshrined in scientific materialism. Throughout the whole period, Whitehead notes, there persists the fixed scientific cosmology which presupposes the ultimate fact of an irreducible brute matter spread through space, whose behaviour was expressed in particular in Newton's laws of motion.

The development of the experimental method and inductive reasoning then occurred as a reaction to the deductive rationalism of medieval thought. For Whitehead, however, the key to the process of induction as assumed either in science or in ordinary life, is not to be found in any Baconian inductive logic but rather in a right understanding of our immediate experience of nature. "Either," he says, "there is something about the immediate occasion which affords knowledge of the past and the future, or we are reduced to utter scepticism as to memory and induction."[13] Whitehead believes that our immediate experience discloses to us that it is included within a larger system of relationships, and that it is from this apprehension of the systematic character of our experience that our faith in the order of nature arises.

Whitehead proceeds to examine the concept of scientific materialism more closely. Its fundamental assumption, he tells us, is that matter has the property of simple location, namely, that it is situated in a definite region of space during a definite lapse of time without reference to other regions of space and other times.[14] Consequently, the answer which the seventeenth century gave to the ancient question of the Ionian thinkers " 'What is the world made of?' was that the world is a succession of instantaneous configurations of matter – or of material."[15]

Another assumption of the philosophy of science of that period is exemplified in the concepts of substance and quality. Substance and quality as well as simple location, Whitehead points out, are among the most natural ideas of the human mind in terms of which we come to order our common sense world. But how concretely, he asks, are we thinking when we consider

[12] *Ibid.*, p. 28.
[13] *Ibid.*, p. 58.
[14] *Ibid.*, p. 64.
[15] *Ibid.*, p. 65.

nature under these assumptions? We are, Whitehead believes, presenting ourselves with simplified editions of matter of fact which are elaborate constructions – examples of what he terms the fallacy of misplaced concreteness. Although it is true, he says, that "as a point of individual psychology we get at the ideas by the rough and ready method of suppressing what appear to be irrelevant details."[16]

Whitehead explains how the notions of substance and quality arise, and here he repeats from a somewhat different standpoint the criticisms of the bifurcation theory which he made earlier in *PNK* and *CN*. We observe, he says, a body which may be hard, blue, round and noisy, and then posit an entity as a substratum (or substance) of which we predicate these qualities. Certain of these qualities are said to be essential: they are permanent features of that entity; others are regarded as accidental, since they can change. The qualities of having a quantitative mass and simple location were held by John Locke to be the essential (or primary) qualities, whilst the accidental qualities were taken by him as qualities of the mind (and therefore secondary) projected on to material bodies.[17] The result of this is that "nature gets credit which should in truth be reserved for ourselves: the rose for its scent: the nightingale for his song: and the sun for his radiance. The poets are entirely mistaken. They should address their lyrics to themselves, and should turn them into odes of self-congratulation on the excellency of the human mind. Nature is a dull affair, soundless, scentless, colourless; merely the hurrying of material, endlessly, meaninglessly."[18]

For Whitehead this paradoxical conception of the universe is framed in terms of high abstractions, elaborated and handled by mathematicians. Their enormous success in enabling us to make predictions about natural phenomena has yielded on the one hand *matter* with its *simple location* in space and time, and on the other hand mind perceiving and suffering and reasoning but not interfering. Modern philosophy, he thinks, has therefore been ruined. It has oscillated between different varieties of monism and dualism. But this juggling with abstractions, Whitehead continues, can never overcome the inherent confusion introduced by the ascription of misplaced concreteness to the scientific scheme of the seventeenth century.[19] Whitehead readily acknowledges the great advantages of limiting thought to clear-cut things which have between them clear-cut relations. If the abstractions are based on important elements of our experience, the scientific thought which

---

[16] *Ibid.*, p. 68.
[17] *Ibid.*, p. 68-69.
[18] *Ibid.*, p. 70.
[19] *Ibid.*, pp. 71-72.

confines itself to these abstractions will arrive at a variety of important truths relating to our experience of nature. But however well-founded be the abstractions to which one limits oneself, one thereby cuts oneself off from that which has been excluded.[20]

Consequently, in the eighteenth century Whitehead tells us, we find a fundamental incoherence in European thought: a scientific realism based on mechanism conjoined with an unwavering belief in the world of man and of higher animals as being made up of self-determining organisms.[21] The discrepancy between the materialist mechanism of science and the moral intuitions which are presupposed in the concrete affairs of life, only gradually manifested its true importance. It was, however, in literature rather than philosophy that the concrete outlook of humanity received its expression.[22]

Whitehead specifically selects Wordsworth as the poet who more than any other openly expressed a moral revulsion against the mentality of the eighteenth century, which accepted the scientific ideas of the period at their face value. He felt that something had been left out of the scientific account of nature and what had been left out comprised everything that was most important for him. Wordsworth opposed to the scientific abstractions his full concrete experience. His consistent theme is, Whitehead tells us, that the important facts of nature elude the abstract analysis of science. The point Whitehead wishes to make in citing Wordsworth is that we forget how strained and paradoxical is the view of nature which modern science imposes on our thought.[23] The English poetic literature of the nineteenth century brings out clearly the radical opposition between the aesthetic intuitions of mankind and the mechanism of science. And Whitehead concludes: "the nature-poetry of the romantic revival was a protest on behalf of the organic view of nature, and also a protest against the exclusion of value from the essence of matter of fact."[24]

In discussing the important advances made by science and technology in the nineteenth century, Whitehead notes four great novel ideas which were introduced during that period.[25] These are: (1) the idea that there is a field of physical activity pervading all space, which presupposes the notion of continuity; (2) the notion of atomicity introduced by John Dalton's atomic theory; (3) the doctrine of the conservation of energy and (4) that

[20] *Ibid.*, pp. 74-75.
[21] *Ibid.*, p. 94.
[22] *Ibid.*, pp. 98-101.
[23] *Ibid.*, pp. 102-103.
[24] *Ibid.*, p. 115.
[25] *Ibid.*, p. 119.

of evolution, these two doctrines being concerned with the notion of transition or change. These developments further undermined the classical conception of materialism, which in particular seems basically incompatible with evolutionary theory. On the materialist theory, Whitehead points out, there is merely change, purposeless and unprogressive. "But the whole point of the modern doctrine is the evolution of the complex organisms from the antecedent states of less complex organisms. The doctrine thus cries aloud for a conception of organism as fundamental for nature."[26]

As a result of these new ideas science, Whitehead believes, is taking on a new aspect which is neither purely physical nor purely biological: it is becoming the study of organisms.[27] What is now needed, he says, is a philosophy of organism to replace the older materialism. This does not mean, as has sometimes been supposed, that Whitehead wishes to base physics on biology, but merely that he believes that they both deal with systems having a historicity about them. He is pointing out that in the physical world as well as in human experience, the way things develop is determined not only by the present situation but by their whole history.

One reason why Whitehead believes that science is nowadays on a higher imaginative level, is that in the past it was closely tied to common sense observations. But with the introduction of better instruments,[28] it has now accumulated a good deal of information about regions of nature far removed from our ordinary experience. In this context he instances Michelson's interferometer experiments, which show that the velocity of light has one and the same speed relative to the apparatus, so that for an instrument on earth and one on a comet, space and time have different meanings. Thus whereas classical materialism presupposed a definite present at which all matter is simultaneously real, on the modern theory there is no unique present situation.[29]

If we turn to the microscopic level we see that quantum theory[30] with its conception of discontinuous energy changes,[31] has also occasioned difficulties for materialism, which assumes the continuity and the simple location of matter. We are now faced, Whitehead points out, with the problem of discontinuous existence and this cannot be understood as long as we explain such energy changes in terms of material particles. Some theory of disconti-

[26] *Ibid.*, p. 130.
[27] *Ibid.*, p. 125.
[28] *Ibid.*, p. 137.
[29] *Ibid.*, pp. 138-142.
[30] *Ibid.*, cf. Chap. VIII, "The Quantum Theory," pp. 153-162.
[31] *Ibid.*, p. 160.

nuous existence becomes necessary. The way out of this difficulty, Whitehead suggests, is to apply to the endurance of matter the same principles as we now accept for sound and light: a steady musical note is the outcome of a vibration in the air, and a steady colour a series of electromagnetic vibrations. In a similar way the system forming the primordial element or organism, requires the whole period in which to manifest itself,[32] so that periodicity is of its very essence. The concept of periodicity is however not new, it is as Whitehead notes, already to be found in the work of Pythagoras who used number to characterise the periodicities of notes.[33]

Whitehead develops in more detail his philosophy of organism when he tells us, that a non-materialist philosophy will identify a primary organism as the emergence of some particular pattern grasped in the unity of a real event. On this view each event involves the pattern of aspects of other events which it grasps into its own unity (its own prehension). Whitehead explains what he has in mind here by quoting an illustrative passage from Berkeley. "*Euphranor*. Is it not plain therefore, that neither the castle, the planet, nor the cloud, *which you see here*, are those real ones which you suppose exist at a distance?"[34] There is, Whitehead continues, a prehension *here* in this place of things which have a reference to other places.[35] Thus the idea of simple location has now gone. The things which are grasped into a realised unity, *here* and *now*, are not the castle, the cloud, and the planet simply in themselves, but aspects of them as observed from the standpoint of our bodily event.[36] Although this account of prehensions is largely concerned with sense-perception, Whitehead believes that similar considerations apply to our descriptions of events in the physical world.

Whitehead claims that as opposed to the Cartesian view which postulates an independent public world and a private world of mind, the effect of physiology was to "put the mind back into nature."[37] Thus in our account of experience it is not now possible to separate mind from the bodily event, which is functioning in nature on the same level as other events. The private psychological field is therefore regarded by Whitehead as merely the bodily event considered from its own standpoint. The primary situation, he says, disclosed in cognitive experience is "ego-object" (or consciousness *here now*) amid other objects within the world of realities.[38]

[32] *Ibid.*, pp. 50-51.
[33] *Ibid.*, p. 52.
[34] *Ibid.*, p. 85.
[35] *Ibid.*, p. 86.
[36] *Ibid.*, p. 87.
[37] *Ibid.*, pp. 173-4.
[38] *Ibid.*, p. 177.

Whitehead concludes *Science and the Modern World* by summarising his position there. He has tried, he tells us, "to analyse the reactions of science in forming that background of instinctive ideas which control the activities of successive generations,"[39] and which permeated the philosophical thought of the period under discussion. In particular he has shown how the concept of the simple location of matter and mind each existing in their own right without reference to each other, has stood in the way of an adequate philosophy of nature. He has himself attempted to state an alternative philosophy in which the concept of organism takes the place of simply located mind and material. On this view mind as involved in the materialist theory, becomes a function of organism.

II. THE REALM OF POSSIBILITY

One of the most interesting as well as puzzling concepts to be found in *SMW*, is that of the "realm of eternal objects," which has been compared to Plato's realm of ideal forms. Whitehead states that our immediate experience is diversified or marked out by reference to a realm of entities which have the possibility of characterising other events in an analogous or different way.[40] For example, a definite shade of orange may be implicated together with a spherical shape in some definite experience. But these two qualities can have other relationships to other events. Instead of manifesting themselves in the orange ball in front of us now, they may on another occasion appear as characterising the setting sun.

Eternal objects are said to be eternal because unlike events they can recur in our experience. Further, as abstract concepts or universals they can be grasped in thought without any actual exemplification in fact.[41] In *SMW* they have a two-fold significance: they can refer (1) to such class concepts as colour or shape, or (2) to certain basic structural relationships which are exhibited concretely in the spatio-temporal relationships existing between events. Whitehead refers to (1) as the "individual essence" of an eternal object and (2) as its "relational essence."[42]

The peculiar and paradoxical nature of Whitehead's realm of eternal objects, which he also describes as the realm of alternative possibilities, is that we are not concerned there with such qualitative characters as red, green,

---

[39] *Ibid.*, p. 224.
[40] *Ibid.*, Chap. X "Abstraction", pp. 184-201.
[41] *Ibid.*, p. 185.
[42] *Ibid.*, p. 186.

etc., (i.e. with their individual essences) but with groups of unspecified objects or patterns of connection in their manifold relationships with each other (i.e. with their relational essence). In his discussion of this realm Whitehead is interested in purely formal structures and not in qualitative detail. Only when these structures are concretely exemplified in our experience in the form of perceptual objects, is there what he terms a real "togetherness of their individual essences."[43]

In his discussion of the realm of eternal objects Whitehead also gives an account of his concepts of being and not-being.[44] These concepts are certainly not new in philosophy. In the *Sophist* Plato already discussed the nature of not-being and in more recent times Sartre in *Being and Nothingness* has used the concept of negation to characterise, among other things, the world of thought and imagination. In a somewhat similar manner Whitehead uses the concept of not-being to refer to eternal objects which are not exemplified concretely in sense-experience, and which can be grasped as possibilities by us in thought. When, for example, we see the red book in front of us now, the eternal object red is said to be included in it as being. But when we go to the window and look out at the green lawn, red then becomes not-being for it – a purely conceptual possibility.

The concept of not-being is closely tied up with Whitehead's belief that the characteristics of our experience might have been otherwise. This he expresses by saying that each occasion of experience is set within a realm of alternative possibilities – the untrue propositions which can be predicated significantly about that occasion.[45] But Whitehead is not simply concerned with false empirical propositions, but with the larger question of the relevance of propositions "disclosed by art, romance, and by criticism in reference to ideals."[46]

Just as the world of our everyday experience can be analysed into a collection of definite perceptual objects, so in the same way Whitehead believes the realm of eternal objects can be analysed into a multiplicity of complex eternal objects of varying grades of complexity. These range from simple eternal objects, such as a definite shade of blue, to more complex eternal objects, such as tables, chairs, trees, etc., conceived as possibilities for thought (i.e. as not-being). The realm of eternal objects would also cover the complex eternal objects which might have marked out a different course of history, as well as objects occurring in the realm of literary creation. The point

[43] *Ibid.*, pp. 186-7.
[44] *Ibid.*, p. 190.
[45] *Ibid.*, p. 185.
[46] *Ibid.*, p. 185.

Whitehead is then making is that despite the interrelatedness of things, we are able to discriminate relatively isolated systems both within perception and in the more abstract realm of possibility.[47]

A complex eternal object having a set of simpler objects subsumed under it is termed by Whitehead an abstractive hierarchy,[48] of which two sorts are distinguished: finite and infinite. An abstractive hierarchy is said to be finite if it includes a limited set of simpler structures, as is the case, for example, in the data given in memory or imagination. It is infinite if it includes objects of all degrees of complexity,[49] as is the case in perception, when we are aware of the distant environment fading away in the general knowledge that there are things beyond.

Whitehead then considers an actual occasion of experience to be "a prehension of one infinite hierarchy (its associated hierarchy) together with various finite hierarchies."[50] Expressing this in somewhat simpler terms, our total experience consists of one such spatio-temporal perspective marked out by a display of interconnected sense-data, together with the various data given in such mental states as thought, memory and imagination. In order to show how these different aspects of our experience may be related to each other, Whitehead introduces a principle which he calls the "Translucency of Realisation."[51] This states that any eternal object in whatever mode of experience it appears, will always retain its self-identity. It is this principle, he tells us, which makes the correspondence theory of truth possible: we can be aware of the same object in different kinds of experience. We can, for example, remember the red we saw yesterday and at the same time check our direct perception of the red we are observing now.

From what Whitehead says about the realm of eternal objects, it may still be difficult to see what sort of ontological status he gives to this realm. It is clearly not a Platonic world of ideas, since it has no independent existence in itself. Nevertheless, eternal objects are not simply subjective creations, they have an objectivity about them – the same self-identical object can be recognised in perception and entertained in thought by different individuals and by the same individual at different times. However, whatever their mode of occurrence, they require on Whitehead's view to be related as elements in

---

[47] For example, in perception. "We endeavour to lift into consciousness meaningful units, such as the whole picture, the whole building, the living animal, the stone, the mountain, the tree." (*Modes of Thought*, pp. 169-70).

[48] *SMW*, pp. 195-6.

[49] *Ibid.*, p. 196.

[50] *Ibid.*, p. 200.

[51] *Ibid.*, p. 200.

certain basic structures. Since in the abstract system of possibilities the qualitative content of eternal objects (i.e. their individual essences) is only indicated ambiguously – one might echo Quine's aphorism and say that for Whitehead "to be" is to be the value of a variable.

CHAPTER 7

THE PHILOSOPHY OF TIME

I. THE PHENOMENOLOGY OF TIME-CONSCIOUSNESS

Whitehead was acutely aware of the problems which arise if one does not want to dismiss the time of our direct human experience as a merely illusory reaction on the part of our minds to the physical world. Although interested in time as a human phenomenon, Whitehead was also concerned to show its relation to the time of scientific thought, which he regarded as only dealing with certain formal relational aspects of our changing human experience.[1]

One finds Whitehead's position echoed among philosophers interested in the phenomenology of temporal awareness. Merleau-Ponty, for example, is highly critical of the abstract concept of time when it is taken as descriptive of the time of human experience. In ordinary life, he points out, "Everyone talks about Time, not as the zoologist talks about the dog or the horse, using these as collective nouns, but using it as a proper noun."[2] Time, he goes on, is sometimes even personified and regarded as "a single, concrete being, wholly present in each of its manifestations, as is a man in each of his spoken words."[3] Merleau-Ponty believes that "There is more truth in mythical personifications of time than in the notion of time considered, in the scientific manner, as a variable of nature in itself, or, in the Kantian manner, as a form ideally separable from its matter."[4]

In this sort of approach Merleau-Ponty is putting forward a diametrically opposed view of time to that normally accepted in scientific discussions,

---

[1] The following books discuss aspects of Whitehead's philosophy of time: William W. Hammerschmidt, *Whitehead's Philosophy of Time*, New York, King's Crown Press, 1947; Robert M. Palter, *Whitehead's Philosophy of Science*, Chicago, University of Chicago Press, Press, 1960.

[2] M. Merleau-Ponty, *Phenomenology of Perception*, translated by Colin Smith, London, Routledge and Kegan Paul, 1962, p. 421.

[3] *Ibid.*, p. 421.

[4] *Ibid.*, p. 422.

one which agrees more with the descriptions given by literary men and artists than that given by physicists. Of course, much depends upon what is truth for whom. Obviously his remarks apply specifically to the human cultural situation rather than to the scientific context, in which the personification of time (as used in the physicist's sense) would be regarded as a gross anthropomorphism. A physicist could argue that a true picture of time can only be arrived at by a study of such things as atomic clocks and other physical processes. The philosopher, historian, poet and artist can only be interlopers in such a field, especially as the individual and his cultural world are in the final analysis contained in the physical world, which can be studied by the precise mathematical and experimental methods of physics. As a final knock-down argument, he could assert that the physical world existed in time before human beings and human consciousness ever appeared on the earth.

The latter point has been taken up by Merleau-Ponty, who asks what could be meant by the statement "that the world existed before any human consciousness."[5] One reply to this might be, he says, "that the earth originally issued from a primitive nebula from which the combination of conditions necessary to life was absent."[6] But as against this he makes the point that "every one of these words, like every equation in physics, presupposes *our* pre-scientific experience of the world,"[7] upon which the meaning of the above statement is based. "Nothing," he says, "will ever bring home to my comprehension what a nebula that no one sees could possibly be. Laplace's nebula is not behind us, at our remote beginnings, but in front of us in our cultural world."[8] One might counter such a remark by saying that one can at least conceive or imagine such a nebula, which is presumably what Laplace did, just as one can conceive multi-dimensional spaces without being able to have sense-awareness of them. But what Merleau-Ponty really means here is that these concepts are not independent of man and the cultural environment in which he finds himself – that they ultimately have to be cashed in terms of human meanings.

Whitehead, at least in his early work, attempts to give an analysis of scientific concepts in terms of actual experience. He would, however, rather base it on the data given in our sense-awareness than on the cultural world as Merleau-Ponty does, although their views have much in common. Whitehead criticised traditional scientific theories because, as he puts it, they give

[5] *Ibid.*, p. 432.
[6] *Ibid.*
[7] *Ibid.*
[8] *Ibid.*

no intelligible account of the meaning of such important physical concepts as "velocity," "momentum" and "stress."[9] Against Whitehead's position one can either put forward something like a Platonism – that physical concepts are intellectually intuited and are hence not reducible to (or derivative from) sense-experience; or accept a conventionalist approach, and say that scientific concepts are postulated and hence do not depend for their meaning on sense-experience, but on our freely created concepts or on our definitions of them. Whitehead's views on this question have a somewhat deceptive simplicity, as he uses some very sophisticated logico-mathematical machinery whose ontological status is not always clear, to relate scientific concepts to empirical data.

Merleau-Ponty, following Husserl, conceives time as a process of "self-production." This he distinguishes from what he calls constituted time, the series of possible relations in terms of before and after and which he claims is not time itself, but the ultimate recording of time.[10] This sort of time has a spatial character since its moments coexist in thought in the form of a linear series. On the other hand, in what he calls true time "with the arrival of every moment its predecessor undergoes a change." Hence, he claims, "time, in our primordial experience of it, is not for us a system of objective positions, through which we pass, but a mobile setting which moves away from us, like the landscape seen through a railway carriage-window."[11]

Interestingly enough a similar position is taken up by the physicist Bridgman, for whom the time of experience consists of a blurred sequence of memories culminating in the budding and unfolding present having a unique apex with the possibility that everything may go awry. He points out that under the influence of the mathematical expression of time in scientific theorising, we nearly all think of time as a homogeneous unlimited one-dimensional sequence, all past time on the one side, all future time on the other, separated by the present which is in continuous motion from the past to the future. Because of this we tend to assume that the future has existence and is essentially predictable. Bridgman contrasts this approach with that of the Greek who thought of himself as facing the past with the future coming up over his shoulder, as the landscape unfolds to one riding back to the engine in a train. Although, he adds, even this picture did not get rid of the idea of the existence of the future, but it did emphasise that the future is unknown.[12]

[9] A.N. Whitehead, *An Enquiry Concerning Natural Knowledge*, cf. Chap. 1, "Meaning."

[10] *Phenomenology of Perception*, cf. p. 415.

[11] *Ibid.*, pp. 419-420.

[12] Percy W. Bridgman, *The Nature of Physical Theory*, Princeton, Princeton University Press, 1936, cf. pp. 29-32.

One of the difficulties in understanding Whitehead's views on time is that of grasping his philosophical approach to experience. His refusal to bifurcate nature is really a rejection of Cartesianism, with its doctrine of psychic additions to nature. He refuses, as we have seen, to divide the data of perception (what he calls the seamless coat of experience) into primary and secondary qualities; the former belonging to the perceived objects, the latter a product of mental excitement. For Whitehead everything perceived is in nature. "We may not pick and choose. For us the red glow of the sunset should be as much part of nature as are the molecules and electric waves by which men of science would explain the phenomenon."[13]

In his early work, at least, he is concerned with describing and analysing how these various elements of nature are connected. In this respect he conceives himself as adopting our immediate instinctive attitude towards perceptual knowledge which is only abandoned under the influence of theory. Whitehead's anti-bifurcationist approach has a certain resemblance to Husserl's phenomenological reduction. Husserl with his watch-word "Back to the things" argues that the analysis of meanings and opinions whether of common-sense or more sophisticated positions, is not the primary objective of philosophy. Philosophy, he argues, must begin with the phenomena themselves – all study of theory takes second place. Thus instead of trying to explain our perceptions by means of physical stimuli and changes in our nervous system, which involve a reference to sophisticated scientific theories, we should concentrate on describing the immediately observed.

In this connection it is of some interest to examine Husserl's phenomenological description of time-consciousness. Husserl argues that we must first completely exclude any assumption, stipulation or conviction concerning objective (or public) time. Just as the objective real world is not a phenomenological datum, so also is not the time of natural science, in which he includes psychology. By objective time he then has in mind that in which all things and events – material things with their physical properties, minds with their mental states – have their definite temporal positions which can be measured by chronometers.

Although objective time might have its basis in immediate experience, Husserl is not primarily concerned with this problem. He makes it clear, for example, as Whitehead does also, that the sensed equality (or simultaneity) of phenomenological temporal intervals cannot be equated with the objective equality of intervals of physical time.[14] Husserl is primarily con-

[13] A.N. Whitehead, *The Concept of Nature*, p. 29.
[14] E. Husserl, *The Phenomenology of Internal Time-Consciousness*, translated by James S. Churchill, The Hague, Martinus Nijhoff, 1964, cf. p. 26.

cerned with an epistemological analysis of temporal lived experience – with its meaning and descriptive content – and not with a causal study of mental states in terms of their development, formation and transformation according to natural laws.

In his discussion of time-consciousness Husserl endeavours to analyse some of the formal relationships which we may discern in our awareness of temporal passage. Examples are: "(1) that the fixed temporal order is that of an infinite, two dimensional series; (2) that two different times can never be conjoint; (3) that their relation is a non-simultaneous one; (4) that there is a transitivity, that to every time belongs an earlier and a later; etc."[15]

His account of temporal experience bears a certain resemblance to that of Whitehead except that the latter's is worded in terms of events rather than temporal intervals, and between which holds the relation of extension which describes the way a larger event extends over smaller events in our perceptual experience. Whitehead's view that public time is of a much more abstract character than experienced time also has certain similarities with Husserl's position. Husserl, unlike Whitehead, is not primarily concerned to give an analysis of the structure of perceived temporal change. Further, since for Whitehead the basic units of experience are events which have both spatial and temporal components, he does not regard time itself as a primitive datum of experience.

II. THE SCIENTIFIC THEORY OF TIME

Whitehead rejects the classical Newtonian theory of absolute time, which is taken as self-subsistent and independent of its content, i.e., matter. On this theory time is regarded as an ordered succession of durationless instants, which are known to us as relata in the serial time-ordering relation. We are aware of this time-ordering relation in itself concurrently with our knowledge of the things occurring at these instants. In holding this view Newton was no doubt influenced by the order of numbers and their generality: points and instants are in this respect like numbers – they are indifferent to whatever is situated in them.

Whitehead also rejects the relational theory of time, which in the history of thought is primarily associated with the name of Leibniz. On this theory time is a set of relations having as relata either material things or sense-qualities. With Hume, for example, time is thought of as a certain order of our impressions conceived as passive in character. Whitehead would argue

[15] *Ibid.*, p. 29.

that in both cases such timeless endurances cannot give us the flux and the creative passage of nature we experience.

Whitehead asserts that we never directly experience time as a succession of instants, and that one can only think of it metaphorically either as a succession of dots on a line or a set of values in certain differential equations. Although the postulation of points as ultimate or primitive elements is quite legitimate for mathematicians engaged in purely mathematical studies, it is another matter if the mathematical point continuum is given a role beyond its analytical one, and regarded as a basis for our description of experienced events. The product of such an analysis, he asserts, is time "as a simple linear series of durationless instants with certain mathematical properties of serial continuity."[16]

The mathematical concept of time has, Whitehead notes, tacitly crept from books on mathematical physics into general scientific thought as expressive of the ultimate structure of space-time. Whitehead claims that on this view velocity, for example, cannot be defined by simple reference to one instant, since one needs a neighbourhood of instants to do this. Further, in the biological field every expression of life takes time, as nothing that is characteristic of life can manifest itself at an instant.

It is clear that Whitehead is right to maintain that we never perceive temporally unextended instants, as all our factual knowledge is confined to observations over a period of time. The belief that there is an instantaneous present directly experienced by us is, he says, a case of warping experience by theory. For Whitehead, what we are immediately aware of is a duration with temporal thickness, its earlier boundary being blurred by a fading into memory and its later boundary by an emergence from anticipation: the present is a wavering breadth of boundary between the two extremes.[17] We are, he says, so trained by language and formal teaching to express our thought in terms of the materialist analysis of time that we tend to ignore the fact that the primary concrete unit discriminated by us in nature is the event retaining its character of passage. Whitehead concludes therefore that the whole conception of nature at an instant is an abstract conception, involving the belief that there is an ideal exactitude of observation, a belief which nevertheless is useful for the purposes of common sense and science.

One of the reasons why Whitehead would not have been sympathetic to the ordinary language approach to philosophical problems and especially to that of experienced time, is that he believes that ordinary language is

---

[16] A.N. Whitehead, "Time, Space and Material," *Proceedings*, Aristotelian Society, Supplementary Volume II, 1919, p. 44.

[17] A.N. Whitehead, *The Concept of Nature*, cf. p. 69.

designed to express clear-cut concepts and that not all sensed phenomena fall within the simplified classificatory criteria of ordinary language. He readily admits that the set of abstract concepts implicit in the structure of ordinary language (including the linear concept of time) has proved itself to be of great pragmatic value in enabling us to handle our common sense world. Nevertheless, Whitehead believes that such language only gives us a useful abstract for the purposes of life.[18]

Whitehead's attempt to ground the concepts of science on our experience may be contrasted with the views of someone like Northrop who is strongly critical of Whitehead's anti-bifurcationist programme. Science, Northrop argues, has to admit a difference between the postulated and the sensed, "the objective world and the events defined in terms of the scientific objects which compose it are not known by observation but by trial and error postulation, confirmed only indirectly through ... the epistemic correlation of some of its events with the phenomenal events which are immediately sensed."[19] A somewhat similar position has been taken up by Einstein who has remarked that the relation between scientific concepts and sensations "is not like that of soup to the beef, but rather like that of the coakroom check number to the coat."[20]

Concepts for Einstein are then logically independent of sense-awareness, and he believes that "we can view only as miraculous that our sense-experience can be unified by our freely created concepts."[21] For Whitehead, however, the age of miracles is past and he would not be entirely sympathetic with the view that scientific concepts are freely created by pure thought alone divorced from our actual perceptions. Whitehead is concerned with the basic question: why do our abstract scientific concepts expressed in mathematical terms apply so well to the physical world? He would, of course, be the first to admit that this effort to harmonise thought and perception is largely a process of progressive approximation.

[18] A.N. Whitehead, *Process and Reality*, cf. p. 234.
[19] Filmer S.C. Northrop, "Whitehead's Philosophy of Science," in *The Philosophy of Alfred North Whitehead*, edited by P. A. Schilpp, Evanston and Chicago, Northwestern University Press, 1941.
[20] A. Einstein, "Physik und Realität," *Journal of the Franklin Institute*, CCXXI (1936), p. 317. Quoted from Robert M. Palter, *Whitehead's Philosophy of Science*, p. 4.
[21] *Ibid.*, p. 4n.

III. CONGRUENCE AND SIMULTANEITY

An interesting feature of Whitehead's account of space and time is his discussion of congruence or the recognition of sameness in nature, as seen, for
example, in our judgments relating to matching, etc. The recognition of
sameness forms for him the basis of spatial and temporal measurement, and
there is a close connection here with his view that experienced nature gives
us some evidence that events exhibit some uniformity in their relationships
to other events. Russell, Whitehead says, pointed out that apart from minor
inexactitudes a determinate congruence relation is among the factors of
nature which our sense-awareness posits for us. On the other hand, Poincaré
argued that it is doubtful whether there is a factor in nature which might
lead any particular congruence relation to play a pre-eminent role. He
therefore claimed that we were free to choose any criterion of congruence
we wished.[22]

Whitehead sees no answer to either of these contentions if one accepts
a materialistic theory of nature, as on this theory nature at an instant
is an independent fact. If we have to look for our pre-eminent congruence
relation amid nature in instantaneous space, Poincaré, he goes on, is right in
saying that nature on this hypothesis gives us no help in finding it. Since
absolute space and time are no longer acceptable, conventionalism seems to
be the one remaining alternative. It needs to be noted, therefore, that Whitehead is fully aware of the fact that if we start with the concept of instantaneous space, then congruence becomes a purely conventional matter.

But Whitehead believes Russell to be in an equally strong position when
he asserts that as a fact of duration, we do find it and it is a remarkable fact,
Whitehead says, that all mankind without any assignable reason should
agree on fixing attention on just one congruence relation. As to the nature of
this relation, this is to be found in our ability to recognise the persistence of
objects or qualities throughout a period of time. Examples of this are to be
found in such material objects as a measuring rod or a pendulum, as well
as certain sets of physical conditions, such as the uniformity of the conditions
for the uniform transmission of light, which is presupposed by Einstein's
definition of simultaneity.

Such judgments of constancy, Whitehead recognises, are of course not
incorrigible. In practice we try to correct the readings of our measuring
rods or clocks for effects of temperature, gravity, etc., rather than rely on
direct perceptions of self-congruence from one instance of the use of the

---

[22] *The Concept of Nature*, cf. pp. 121-124. *PNK*, Chap. Congruence, pp. 49-50.

measuring rod to another. On the other hand, he points out, even such corrections depend in the last analysis on our judgments of constancy.

The conventionalist defines congruence by the requirement that Newton's physical laws of motion are true. As against this position Whitehead argues that uniformity in change was directly perceived and the measurement of time was known to all civilised nations before Newton's laws were thought of: it is this time as thus measured that the laws are concerned with.[23] As far as judgments of spatial congruence are concerned, it is Whitehead says, a fact of nature that a distance of thirty miles is a long walk for anyone, and this does not seem to be a matter of convention. The process of measurement, he concludes, is merely a procedure to extend the recognition of congruence to cases where these immediate judgments are not available.

Whitehead's views on congruence have been criticised by Grünbaum. As the latter bases his criticisms on an earlier critique of Northrop, concerned with Whitehead's approach to the related question of simultaneity, we need first to examine Northrop's objections. Northrop argues that Whitehead's view that we have a direct awareness of simultaneity, conflicts with the basic definition of the temporal relation of simultaneity for spatially separated events in terms of light waves upon which definition the theory of relativity rests. Both common sense and Einstein's physics, he goes on, when they admit a public time the same for all observers, bifurcate nature into the intuited relation of simultaneity varying from person to person, and the postulated simultaneity of physical theory, the same from person to person at rest relative to each other on the same frame of reference.[24] Northrop therefore concludes that science requires bifurcation and indeed cannot avoid it.

Northrop, however, fails to appreciate that Whitehead is concerned with sense-experience, before one applies the categories public/private to it. Our immediate experience for Whitehead has a subject-object structure in which we are aware of ourselves as related to other perceived events in nature, between which a relation of simultaneity holds. Although Whitehead believes there is an objective world of events which we share in common with others, and of which different percipients may have different perspectives, this public world of common sense already involves an element of interpretation for him. He would not deny that science requires bifurcation, and he would readily admit that science would not have advanced without it. What he objects to is taking the scientist's system of abstract concepts as valid for the rest of experience.

[23] *The Principle of Relativity*, pp. 49-50.
[24] Filmer S.C. Northrop: "Whitehead's Philosophy of Science," cf. pp. 200-201.

In any case Whitehead distinguishes just as clearly as does Northrop, between perceived simultaneity and what he calls instantaneousness (or nature at an instant) which forms the datum of science and corresponds to Einstein's concept of simultaneity. But Whitehead's concept of simultaneity is not concerned with light or any other type of physical signal and applies only to actual perceivable events. Furthermore, Whitehead makes it clear that in a physical system employing the concept of instantaneousness, no such unique relation will exist. To some extent then, the conflict between Whitehead's account of simultaneity and that of the relativity physicist is a verbal one, but not entirely, as he would regard the public time of the astronomer which Northrop falls back upon, as involving a large degree of conceptual interpretation.

Grünbaum has taken up some of the points made by Northrop. He regards the concept of congruence – our means of judging spatial or temporal equalities – as conventional. With regard to simultaneity, Grünbaum points out, Einstein assumes that within the class of physical events the readings of natural clocks do not define relations of absolute simultaneity under transport (i.e. they are non-synchronous). The failure of human signalling and measuring operations to disclose relations of absolute simultaneity is therefore only the epistemic consequence of the primary non-existence of these relations.[25]

Grünbaum does not think that Whitehead's historical observation that the human race possessed a metric prior to the statement of Newton's laws invalidates Poincaré's contentions "that (1) time-congruence in physics is conventional, (2) the definition of temporal congruence used in refined physical theory is given by Newton's laws and (3) we have no direct intuition of the temporal congruence of non-adjacent time-intervals."[26]

In the case of space, Grünbaum would claim that it too is deficient in such a metric. He does not therefore think that Whitehead is entitled to regard coincidence (or matching) as the only test of congruence; that measurement presupposes a criterion does not at all invalidate the conventionality of the self-congruence of our rods at different places. He tries to meet Whitehead's point that a distance of thirty miles is a long walk for anyone, by saying that this is due to our gait being tied to the yard (or metre) stick, "thus making it an objective fact that an interval which measures thirty miles in the metric

[25] A. Grünbaum, "Whitehead's Philosophy of Science," *The Philosophical Review*, LXXI (1962), pp. 222-223.

[26] A. Grünbaum, *Philosophical Problems of Space and Time*, New York, Knopf, 1963, p. 53.

of the yardstick will contain a great many of our steps."[27] But he does not believe that this shows that the self-congruence of the yardstick is non-conventional. What Grünbaum seems to overlook here, is the feeling of fatigue that a normal person would have after he had walked a distance of some thirty miles. In most cases fatigue feelings are a direct function of the number of steps taken, but as a phenomenological fact one usually experiences the fatigue without having counted the steps, unless one is the sort of obsessional who is careful not to step on the cracks of pavements.

Grünbaum further argues that the agreement which obtains between the metric of psychological time and the physical time congruence defined by Newtonian laws and a variety of physical processes, cannot be invoked against the conventionalist view. The ability of men and animals to make successful estimates of duration derives from the fact that the metric of psychological time is tied causally to the physical processes which may define time congruence in physics. "How then," he asks, "can the metrical deliverances of psychological time possibly show that the time of physics possesses an intrinsic metric, if, as we saw, no such conclusion was demonstrable on the basis of the cycles of physical clocks?"[28]

Grünbaum, unlike Northrop, does not sufficiently realize that Whitehead is talking about sensed simultaneity and not instantaneousness (or simultaneity in physics). This is particularly the case when Grünbaum quotes experimental work from biology and psychology to show that psychological time as investigated empirically (in terms of chronometers, etc.) is merely a function of physical time and is hence no guide to it. Grünbaum would seem to be appealing to something like a materialistic theory of perception as evidence for the falsity of Whitehead's views. But such a theory for Whitehead gives what he considers to be abstract scientific notions a greater reality than the perceptually given. Furthermore, psychological time as investigated by the psychologist in his laboratory would for Whitehead be just as much an interpretation of experienced time as are the accounts of time given by the physicist.

In any case Whitehead is not primarily concerned with the causes of our knowledge. He is engaged in an epistemological rather than a causal enquiry. Whitehead would say that the congruence judgments of physicists, biologists and psychologists, ultimately depend on their recognition of self-identities in their experience of matching, etc.

Commenting on Whitehead's anti-bifurcationist philosophy of nature, Grünbaum says that the verdict on it must be the same as that which Pauli

[27] *Ibid.*, p. 62.
[28] *Ibid.*, cf. p. 60.

gave on unified field theory, namely, "What God hath put usunder no man shall join!"[29] For Whitehead as far as our immediate experience is concerned, the reverse is true.

IV. DISCRETENESS AND CONTINUITY IN TIME

In a more recent work,[30] Grünbaum tells us that James and Whitehead, with Zeno's paradoxes of the Dichotomy and Achilles in mind, believe that time does not possess the structure of the linear mathematical continuum. According to Grünbaum their reasons for this denial are: "(1) the relations of temporal order among physical events *as they actually happen* are as known to us in our conscious awareness of their coming into being; (2) occurring *now* or happening is pulsational and not punctual" (i.e. our perceptions have a temporal threshold) "and (3) the serial order of pulsational coming into being is *not* dense but *discrete*."[31] If we assumed, he goes on, that the temporal order of events is isomorphic with the discrete order of nows of awareness, we should have to deny that the serial order of time is dense.[32]

Grünbaum attempts to justify the denseness of physical time (i.e. its infinite divisibility) by denying that physical events as such come into being at all – that they merely occur tenselessly in a network of timeless separation. He therefore rejects the view that physical events must occur in the pulsational consecutive manner in which they are perceived to come into being. The coming into being of events is rather regarded by him as a mind-dependent quality. "Nowness" and "temporal becoming" are like colour and taste not properties of physical events, and have no existence apart from our minds.[33] Hence, he believes that perceived temporal change has no counterpart in the time of physics. The atomicity of becoming, he concludes, "thus turns out to be expressive of an organismic feature of our kind of nervous system instead of warranting the quantization of physical time."[34]

Although it is true that Whitehead uses such words as "now," "here," etc., to refer to the way we are aware of ourselves as related to the rest of our

[29] A. Grünbaum, "Whitehead's Philosophy of Science," p. 229.

[30] A. Grünbaum, *Modern Science and Zeno's Paradoxes*, London, Allen and Unwin, 1968.

[31] *Ibid.*, pp. 45-46.

[32] *Ibid.*, cf. p. 52.

[33] *Ibid.*, cf. p. 55.

[34] *Ibid.*, p. 56.

perceptual field, this is an entirely different relationship from the one involved in the theory of mind-dependence of colours and tastes, a theory which
is concerned with our mental reaction to physical stimuli. Presumably in
the same sort of way "temporal becoming" is to be regarded as a psychological reaction to the tenseless seriality of physical time. Grünbaum's account of temporal becoming is therefore a complete inversion of Whitehead's position, since Grünbaum accepts the view that the structure of the
abstract mathematical continuum is isomorphic with that of public physical
time, and consigns the quality of temporal becoming to the field of subjective
illusion. Indeed the network of timeless separation which Grünbaum takes,
as the very essence of time would seem to be time as contemplated by God
sub specie aeternitatis.

Grünbaum's argument relating to the inability of our nervous system to
perceive continuity in the Cantorian dense sense (i.e. as infinitely divisible),
overlooks the fairly obvious empirical finding that apart from physical
quanta neural impulses are themselves pulsational. Whitehead would argue
that as soon as we start talking of the biological events in our nervous system,
we already presuppose a physiological theory involving the application of
concepts and criteria, which involve procedures of intellectual interpretation.
Whatever James' position might have been, Whitehead has always asserted
that physical time only involves certain features of perceived temporal change,
which we rationally reconstruct in terms of an irreversible temporal series.

Grünbaum would seem to believe that there is broad inductive evidence
that space and time are continuous in the Cantorian sense. Whitehead has,
however, pointed out that owing to the inexactness of measurement, it is
impossible to tell whether a continuous physical quantity possesses the
compactness of the series of rationals or the continuity of the series of real
numbers. Although we usually adopt the latter hypothesis because of its
mathematical simplicity, the assumption has no *a priori* grounds in its
favour. The continuity of space (and time) therefore rests according to
Whitehead upon an assumption unsupported by any *a priori* or experimental
grounds. "Thus," he concludes, "the current applications of mathematics
to the analysis of phenomena can be justified by no a priori necessity."[35]

---

[35] *Essays in Science and Philosophy*, p. 281. Another thinker who denied the existence of
physical passage and who is quoted approvingly by Grünbaum was Hermann Weyl. Weyl
stated, "The objective world simply *is*, it does not *happen*. Only to the gaze of my consciousness, crawling upward along the life-line of my body, does a section of the world come to
life as a fleeting image in space which continuously changes in time" (*Philosophy of
Mathematics and Natural Science*, Princeton, 1949, p. 116). Weyl's position would
seem to be an updated version of Plato's conception of time as a moving image of Eternity.

PART III

METAPHYSICS

CHAPTER 8

PROCESS AND REALITY

I. THE PHILOSOPHICAL SYSTEM

In discussing the philosophical system put forward in *PR*, we need to point out that not only does Whitehead introduce a novel terminology, but the work itself is somewhat amorphous in character, and this despite his attempt to state a categoreal scheme – a general scheme of ideas in terms of which all our experience is to be described. There is also a considerable amount of overlap between the various parts of this book. It might have been a clearer and more effective work if Whitehead had engaged in some judicious pruning before publication.

PR which extends to 497 pages is divided into five parts. In Part I[1] "The Speculative Scheme," Whitehead explains the philosophic method used and gives a summary of the scheme of ideas in terms of which his cosmology is to be framed. In Part II[2] "Discussions and Applications" he tries to exhibit this scheme as adequate for the interpretation of the ideas and problems which form the complex texture of civilised thought. This is followed by a discussion of the philosophical views of the group of 17th and 18th century philosophers and scientists, in particular, Descartes, Newton, Locke, Hume and Kant, which dominated the development of subsequent philosophy; and he brings out the agreements and differences between their philosophical positions and his own. He also discusses some of the more general notions of physics and biology insofar as they are related to the doctrines of the philosophy of organism.

In Parts III[3] and IV,[4] Whitehead tells us in his Preface,[5] the cosmological scheme is developed in terms of its own cosmological notions and without

---

[1] *PR*, pp. 3-50.
[2] *PR*, pp. 53-305
[3] *PR*, pp. 309-97.
[4] *PR*, pp. 401-474.
[5] *PR*, p. vi.

regard to other systems of thought. Part III "The Theory of Prehensions" is largely concerned with giving an account of the way actual entities or events prehend or influence each other in the world. He introduces his view that physical events in their prehensions of each other exhibit a primitive "feeling" character, and goes on to discuss the nature of conscious perception and to develop a theory of judgment.

Part IV "The Theory of Extension" is of a more technical nature, and examines certain basic logical and mathematical notions such as the "Extensive Continuum." Whitehead also gives an updated version of the *Method of Extensive Abstraction*, in which points, lines, etc., are derived from the relation of extensive connection holding between abstract regions, and which is more general than his earlier part and whole relation of extension which held between events. Among other topics discussed is the connection between geometry and perception and the nature of measurement. Part V[6] "Final Interpretation" is largely taken up with his account of the relationship between God and the World. As he says, it "is concerned with the final interpretation of the ultimate ways in which the cosmological problem is to be conceived."[7]

Whitehead's philosophical position in *PR* is then very different from the current conception of philosophy as the analysis of our knowledge and language about the world. He is not simply concerned with making our ideas clear about these questions, especially as he does not believe that our common sense notions can give us an adequate account of reality, however much we may clarify them. His philosophical approach may seem unfashionable today, and it was so already at the time of writing *PR*. He is attempting to construct a cosmology, and this is reflected in the full title of his work *Process and Reality: An Essay in Cosmology*. It is Whitehead's belief that the movement of historical and philosophical criticism which has on the whole dominated the last two centuries, has done its work and requires to be supplemented by an effort of constructive thought.

Whitehead argues that all constructive thought in the special topics of scientific interest is dominated by some philosophic scheme, unacknowledged but no less influential in guiding the imagination. The importance of philosophy, he therefore believes, lies in its sustained effort to make such schemes explicit and capable of criticism and improvement. He singles out the two cosmologies which at different periods have dominated European thought: that contained in Plato's *Timaeus* and the cosmology of the 17th century, whose chief authors were Galileo, Newton and Locke. In

---

[6] *PR*, pp. 477-497.
[7] *PR*, p. vii.

attempting a similar enterprise he thinks that perhaps the true solution is a fusion of these earlier schemes, but with due modifications so as to achieve self-consistency and to take account of the advance of knowledge.[8]

Whitehead is therefore interested in bringing out the limitations of the categorial frameworks implicit in past cosmologies, and in constructing an alternative scheme which will fit in better with our present day scientific knowledge, and at the same time take account of the rich variety of our experience. It must, he says, "be one of the motives of a complete cosmology, to construct a system of ideas which brings the aesthetic, moral, and religious interests into relation with those concepts of the world which have their origin in natural science."[9] However, when one looks more closely into the system of ideas which is to bring together every aspect of our experience, it will be seen that it is in no way *a priori*, but is directly rooted in our experience. Whitehead's cosmological scheme therefore seems to have more of a descriptive than a speculative character.

To aid our understanding of his philosophical system Whitehead gives us the following list of prevalent habits of thought, all of which he repudiates as far as concerns their influence on philosophy.

(1) The distrust of speculative philosophy
(2) The trust in language as an adequate expression of propositions
(3) The mode of philosophic thought which implies, and is implied by faculty psychology
(4) The subject-predicate form of expression
(5) The sensationalist doctrine of perception
(6) The doctrine of vacuous actuality
(7) The Kantian principle of the objective world as a theoretical construct from purely subjective thinking
(8) Arbitrary deductions in *ex absurdo* arguments
(9) Belief that logical inconsistencies can indicate anything else than some antecedent errors.

Whitehead goes on to state that much of 19th century philosophy excludes itself from relevance to the ordinary stubborn facts of life, because of its ready acceptance of some or all of these nine myths and fallacious procedures.[10]

Some discussion is needed of (3), (6), (8) and (9), which may not seem very self-explanatory. In the case of (3) Whitehead contends that philosophical

---

[8] *PR*, cf. p. ix.
[9] *PR*, p. vi.
[10] *PR*, p. viii.

thought has made difficulties for itself by dealing exclusively in very abstract notions, such as those of "mere awareness, mere private sensation, mere emotion, mere purpose, mere appearance, mere causation."[11] These notions, he says, are the ghosts of the old faculties, banished from psychology but still haunting metaphysics. They have, however, no existence in themselves, but always form elements in the experience of some individual subject existing here and now.

In (6) Whitehead is concerned to criticise "the notion of vacuous material existence, with passive endurance, with primary individual attributes,"[12] namely, the view that we can talk significantly about entities existing in nature independently of a subject's experience of them. As he tells us, "apart from the experiences of subjects there is nothing, nothing, nothing, bare nothingness."[13] Entities as they occur in nature are always related to experiencing subjects, either as they are prehended on a low grade experiential level by other events, or on a higher level by the percipient event in conscious sense-perception.

In (8) and (9) Whitehead is pointing out that just because it is possible to show that a philosophical system can give rise to contradictions, this does not mean that it can therefore be simply written off as philosophically valueless. Logical consistency is not the only test of the truth of such a scheme, it must also be shown to be in conflict with the experienced facts. As he notes, "the only logical conclusion to be drawn, when a contradiction issues from a train of reasoning, is that at least one of the premises involved in the inference is false."[14]

II. THE CATEGOREAL SCHEME[15]

In *PR*, Chapter II "The Categoreal Scheme," Whitehead gives an anticipatory sketch of the primary notions which constitute the philosophy of organism. This he does by stating four kinds of category.[16]

    I. The Category of the Ultimate
    II. Categories of Existence
    III. Categories of Explanation
    IV. Categoreal Obligations

[11] *PR*, p. 24.
[12] *PR*, p. 438.
[13] *PR*, p. 234.
[14] *PR*, p. 10.
[15] *PR*, pp. 24-41.
[16] *PR*, p. 27.

The above are not categories in the sense of *a priori* forms of thought impressed upon the material given in sense-experience, but are rather principles which are manifested objectively in "the becoming, being and relatedness of events." They also differ among themselves: whereas I and IV seem to have a more general metaphysical character, II is largely a classification of the basic elements discriminated by us in experience, and III explains in more detail the character and functioning of these entities.

From the way this categoreal scheme is stated one might get the impression that Whitehead is putting forward something like an axiom system, where starting from primitive notions one can construct a coherent system of thought which can then be tested against the facts. Although this would seem to be the ideal he sets himself in his account of philosophical method, where he enumerates the criteria which a successful philosophical scheme should satisfy, this seems far from being the case in actual practice.

What Whitehead seems to be really giving in Chapter II "The Categoreal Scheme," is a summary of the philosophical system he is to elaborate in the later pages of *PR*. As he says, "the whole of the subsequent discussion in these lectures has the purpose of rendering this summary intelligible."[17] If this chapter had been placed at the end of *PR* rather than at the beginning, it might not have given the impression that Whitehead was constructing a categorial framework from which his philosophy proceeded step by step.

I. The Category of the Ultimate[18]

The Category of the Ultimate presupposed in all the other special categories is made up of the three notions of "creativity," "many" and "one." These notions, Whitehead tells us, are at the root of the synonymous terms "things," "being" and "entity." Creativity, which is a principle of novelty, is "that ultimate principle by which the many, which are the universe disjunctively, become the one actual occasion which is the universe conjunctively ... creating a novel entity other than the entities given in disjunction."

It is perhaps unfortunate that Whitehead describes the creation of the novel event, as it arises from antecedent events as the "advance from disjunction to conjunction." In logic, for example, propositions stated in the normal disjunctive form can be translated into the conjunctive normal form

---

[17] *PR*, p. 24. It should be noted that Whitehead uses the phrase "categoreal scheme" rather than the more usual "categorial scheme." And he gives no reason for this spelling variation. One can only assume that he wishes to differentiate his scheme of categories from that of Kant, as his categories are not *a priori* but tentative formulations of the basic metaphysical notions.

[18] *PR*, pp. 28-29.

by negating them according to the De Morgan laws, so that we obtain what are in effect equivalent propositions. In this case, however, the principle involved is a radically different one. It is a Gestalt principle, namely, that "the whole is more than the sum of its parts." It therefore becomes necessary for Whitehead to introduce the notion of creativity to show how a multiplicity of disjunctive entities – the antecedent states of the universe – can give rise to a novel entity. Whitehead uses the term "concrescence" to refer to this coming together of the aspects of other events into a novel unity, which is the present subject prehending the world from its own point of view.

II. The Categories of Existence[19]

These categories classify the different kinds of entities discriminated by us in experience. They resemble the Aristotelian categories conceived as the most general classes in terms of which things may be classified.
They are:

(i)   Actual Entities (or Actual Occasions)
(ii)   Prehensions or Concrete Facts of Relatedness
(iii)   Nexūs (plural of Nexus) or Public Matters of Fact
(iv)   Subjective Forms or Private Matters of Fact
(v)   Eternal Objects, or Pure Potentials for the Specific Determination of Fact
(vi)   Propositions, or Matters of Fact in Potential Determination
(vii)   Multiplicities, or Pure Disjunctions of Diverse Entities
(viii) Contrasts, or Modes of Synthesis of Entities in one Prehension.

Despite the novel terminology used by Whitehead here, a similar classification is to be found in his philosophy of nature. In discussing the diversification of nature into a number of different kinds of entity, he points out that the type of diversification we make is dependent upon the purpose we have in mind. In PNK[20] he confines himself to five modes of diversification which he considered to be important for scientific theory. These are: (1) events, (2) percipient objects (or conscious egos), (3) sense-objects, (4) perceptual objects, (5) scientific objects. Events and sense-objects become in PR actual entities and eternal objects, whilst nexūs, prehensions and propositions deal with more complex systems of events and their associated objects. Subjective forms and contrasts appear for the first time in PR, and refer to the subject's mental reactions to the experienced data. But what Whitehead had not yet developed in his philosophy of nature, was the view that the physical events

[19] PR, p. 29.
[20] PNK, p. 60.

or actual entities in their prehensions of each other exhibit something like a rudimentary feeling tone. In our discussion of the philosophy of organism, we shall use the terms event and actual entity (or actual occasion) inter-changably, since they refer to similar sorts of entity, except that an event now has not only the character of passage, but also a primitive "feeling" tone.

III. The Categories of Explanation[21]

The categories of explanation of which there are twenty-seven, mainly describe the nature and function of the entities enumerated in the categories of existence. There seems, however, no implicit reason why their number should be just twenty-seven, as some of them cover similar ground.

In Categories (i) to (vi) Whitehead states that the world is a process which is the becoming of actual entities and that during this process the potential unity of many entities – actual and non-actual – acquires the *real* unity of one actual entity: novel prehensions, nexūs, subjective forms, multiplicities and contrasts also arise. Further each entity, actual or non-actual has the potentiality for being an element in a *real* concrescence of many entities into one actuality. He makes the further point that no two actual entities arise from the same self-identical universe: with the creative advance of nature, novel entities are progressively added to the universe. On the other hand, there are no new eternal objects – they are the same for all actual entities.

Category (vii) asserts that eternal objects can only be described in terms of their potentiality for ingression in the becoming of actual entities. This category seems to be merely a restatement of his earlier philosophy of nature position, that the relationship between events and objects is a multi-termed one.

Categories (viii) – (ix) are concerned with the analysis of actual entities. Whitehead states that an actual entity can be analysed (a) in terms of its potentiality for objectification in the becoming of other actual entities (i.e. the way it affects other entities) and (b) in terms of the process which consti-tutes its own becoming. The being of an actual entity is therefore constituted by its becoming and this is termed the principle of process.

Categories (x) – (xii) deal with prehensions. A prehension (or a concrete fact of relatedness) can, we are told, be analysed into (a) the subject which is prehending (the notion of a subject here seems to be a generalisation of his earlier notion of a percipient event), (b) the datum which is prehended (e.g. the perceived perspective), and (c) the subjective form, which is how that subject prehends that datum. Prehensions of actual entities are termed

[21] *PR*, pp. 30-35.

physical prehensions, and deal with the way events influence each other, whilst prehensions of eternal objects which refer to the ingression of objects into events, are termed conceptual prehensions. Whitehead notes that consciousness is not necessarily involved in the subjective form of either prehension. On Whitehead's view the causal relations between physical events are describable in terms of non-conscious primitive "feelings" (or physical prehensions), whilst the eternal objects implicated at these levels (presumably as scientific objects) would also occur on a non-conscious level. Whitehead further distinguishes between positive and negative prehensions. The former cover those entities included in the experienced perspective, and the latter, elements (actual and potential) which are excluded from it.

Category (xiii) lists the various types of subjective forms: i.e., the subject's reaction to the perceived data, of which there are many types: emotions, valuations, purposes, adversions, aversions and consciousness. It may seem strange that consciousness is listed as a subjective form, but Whitehead believes that the philosophy of organism has abolished the detached mind. "Mental activity is one of the modes of feeling belonging to all actual entities in some degree, but only amounting to conscious intellectuality in some actual entities."[22]

In Category (xiv) the concept of a nexus is discussed. It covers (1) Whitehead's earlier concept of a duration as when, he tells us, "The nexus of actual entities in the universe correlate to a concrescence, is termed 'the actual world' correlate to that concrescence,"[23] which resembles the associated duration of a percipient event. It also covers (2) the historical routes of events which on Whitehead's view form enduring objects, such as, for example, The Great Pyramid.

Category (xv) is concerned with the nature of propositions. A proposition for Whitehead is made up of a set of actual entities having a predicative pattern of eternal objects. On a conscious perceptual level, the notion of a proposition refers to a set of sense-qualities having a specific spatio-temporal position, for example, a coloured patch at a particular moment of time. In such mental states as imagination or memory similar structures occur in the imaged or remembered data. Whitehead is obviously using the term proposition here in a way that is very different from that in which it is normally used in logic. He believes that the ordinary logical account of propositions expresses only a restricted aspect of their role namely when they, form the data of judgments. The primary function of a proposition, Whitehead says, is to be relevant as a lure for feeling, as when, for example, we

[22] *PR*, p. 77.
[23] *PR*, p. 30.

enjoy a joke. Other propositions, he goes on, are felt with feelings whose subjective forms are horror, disgust or indignation.[24]

Categories (xvi) and (xxvii) cover multiplicities and contrasts. The former notion refers to a group of entities such that, as he puts it, "all its constituent entities severally satisfy at least one condition which no other entity satisfies,"[25] for example, they may all be events in a specific duration. The latter refers to the way different kinds of entity, events and eternal objects, are integrated together in one experience. An example of this is to be seen in what he terms an affirmation-negation contrast, namely, our clear-cut perception of an object, say the house in front of us as standing out against our background of bodily awareness. By introducing the concept of a contrast Whitehead has to admit "that there are an endless number of categories of existence, since the synthesis of entities into a contrast in general produces a new existential type."[26]

Category (xviii) expresses what he terms the ontological principle: that the reasons for the becoming of some actual entity in the actual world lie "*either* in the character of some actual entity in the actual world of that concrescence, *or* in the character of the subject which is in process of concrescence."[27] Whitehead believes that this principle "broadens and extends a general principle laid down by John Locke in his *Essay* (II, XXIII, § 7), when he asserts that "power" is a great part of our complex ideas of substances."[28] Whitehead tells us that in his own account the notion of substance becomes transformed into that of actual entity, and the notion of "power" into the principle that the reasons for things are always to be found in the composite nature of actual entities.

What Locke meant by power may be seen from his statement (*Essay* II, XXI, I). "Thus we say fire has a power to melt gold; ... and gold has a power to be melted: ... Power thus considered is twofold; viz. as able to make, or able to receive, any change: the one may be called 'active,' and the other 'passive,' power ... And sensible qualities, as colours and smells, etc., what are they but the powers of different bodies in relation to our perception!"[29] Whitehead points out that the problem of perception and power are one and the same, at least insofar as perception is reduced to a mere prehension of actual entities, i.e., as far as we refer to the way physical events in-

[24] *PR*, pp. 33-34.
[25] *PR*, p. 32.
[26] *PR*, p. 32.
[27] *PR*, p. 33.
[28] *PR*, p. 25.
[29] *PR*, p. 79.

fluence each other. In the higher grades of conscious perception such "powers" would, as Locke suggests, manifest themselves through our awareness of sense-qualities.

In Category (xix) Whitehead makes it clear that the fundamental types of entity in the philosophy of organism are actual entities and eternal objects, "that the other types of entities only express how all entities of the two fundamental types are in community with each other, in the actual world."[30] His position here is then similar to that expressed in his nature philosophy, where events and sense-objects are taken as being the basic elements given in our experience of nature, all other entities, of which there can be an indefinite number of different types, being regarded as sub-varieties of these two basic kinds.

Categories (xx) – (xxvii) deal with the way actual entities are objectified in other actual entities in the world (i.e., with the way they prehend or influence each other) and with the ingression of eternal objects into actual entities. He also discusses the various phases in the concrescence or the drawing together of aspects of other events into a unified perspective by a specific subject. Its final completed phase is termed by him the "satisfaction."

IV. The Categoreal Obligations[31]
Whitehead elaborates nine categoreal obligations. The first three categories seem to be general principles determining the functioning of all actual entities. The next four refer to processes governing perception at various levels. And the last two are general principles concerned with the determination of the past and the openness of the future. Whitehead's reason for postulating these categories stems from his belief that the order we find in our experience of the world is not due to some intellectual categorisation, but is an exemplification of certain objective principles to which physical and psychological processes have to conform.

(i) *The Category of Subjective Unity*. This deals with the unity given to the various aspects of other events in the universe, as a result of the present event or subject integrating them into its specific perspective. (ii) *The Category of Objective Identity* is concerned with the self-identity of any element in the integrated unity, whilst (iii) *The Category of Objective Diversity* deals with the diversity of these self-identical elements, so that no one entity can take on another's function. Whitehead would argue that these categories have a metaphysical generality about them: that the perceived data are always prehended by a subject which gives them their unity, and that the ele-

---

[30] *PR*, p. 34.
[31] *PR*, pp. 35-38.

ments in the perceived perspective always retain their self-identity and diversity.

(iv) *The Category of Conceptual Valuation* and (v) *The Category of Conceptual Reversion*. On the level of conscious perception, the former category refers to the way in which eternal objects (or sense-qualities) are derivative from the physical events which are said to be causal of them. As we have seen the relation of ingression of objects into events is a multi-termed one, and includes not only a bodily event, but also the active conditioning events in nature. At the lower organic levels in nature the latter category covers the origination of conceptual novelty in living organisms (for example, the appearance of life and the higher degrees of mentality). In conscious sense-perception it is concerned with the way the subject imaginatively originates sense-qualities to reinforce his direct experience. Thus when we are aware of some definite perceptual object, say a horse, it is more than the brown patch of colour immediately given to us. We, as it were, complete it by originating further sense-qualities ourselves, so that we are aware of some definite object, in this case a horse.

(vi) *The Category of Transmutation* covers the process whereby a single enduring object is substituted for the historical route of atomic events (or nexus) occurring in nature. On the conscious perceptual level, for example, we become aware of an enduring object rather than a succession of transitory events. "We have," Whitehead says, "to account for the substitution of one nexus in place of its component actual entities."[32] In the *Monadology* Leibniz, he points out, "meets the same difficulty by a theory of confused perception."[33]

(vii) *The Category of Subjective Harmony* deals with the way the various eternal objects in the perceived perspective are combined together by the subject so as to form a harmonious pattern.

(viii) *The Category of Subjective Intensity* and (ix) *The Category of Freedom and Determination*, are both concerned with the relation of the present event to the future. It is worth looking at these two categories in some detail. *The Category of Subjective Intensity* has to do with what Whitehead terms the "subjective aim" of the subject which is expressed, he tells us, as an intensity of feeling in the immediate subject and relevant future.[34] By the relevant future Whitehead is referring to the subject's anticipation of the future. He believes that our experience of ourselves as modifying our future through our decisions, is the foundation of our experience of respon-

---

[32] *PR*, p. 355.
[33] *PR*, p. 37.
[34] *PR*, p. 37.

sibility and freedom. This element in our experience is, he argues, too large to be put aside merely as a misconstruction, since the whole tone of human life is governed by it. Although it can only be illustrated by striking examples from fact and fiction, these instances are only conspicuous illustrations of human experience during each hour and minute. In other words, our experience of freedom is best seen during critical periods of our existence. Whitehead gives the example of Galileo and the Inquisition. Galileo, he says despite his public recantation was still free to believe that the earth moved.[35]

Whitehead asserts that such freedom from rigid determination by antecedent states of the universe (of the sort presupposed by classical determinism) can also be observed when we come to study physical events in nature. This does not mean that he believes that physical events exhibit "foresight" in their functioning, nor is he trying to justify human freedom in terms of the physical principle of indeterminacy. He is rather seeking to relate our experience of living in a contingent world, with the way in which physical and biological phenomena exhibit an indeterminacy as far as their future states are concerned.

When we endeavour to understand past events we are faced by a closed problem, as the events have already occurred. On the other hand, when we come to deal with the future the problem still remains an open one. Whitehead makes this point in *The Category of Freedom and Determination*. "The concrescence of each individual actual entity is internally determined and externally free."[36] He explains this by saying that the peculiarity of history can be rationalised by the consideration of the determination of successors and antecedents – that in every pattern of history we can trace certain events which have arisen from events in the relevant past. But however effective the concept of efficient causation may be, it seems to break down at the utmost point of unfolding immediacy. Whitehead therefore argues that the evolution of history is incapable of rationalisation as far as the pattern it exhibits is concerned, as no reason internal to history can be found to explain why that course of history rather than any other should have occurred.[37] He would seem to be rejecting here the idea that there is some inner dialectic in history which determines beforehand what types of pattern (or flux of forms) it will exhibit.

The principle that the process of an actual occasion is internally determined and externally free may be summed up as follows. Each event can be said to be internally determined by the events in its immediate past, since

[35] *PR*, p. 64.
[36] *PR*, p. 37.
[37] *PR*, p. 64.

they stand in relations to each other of successor and antecedent. But on the other hand, the specific character that the event will assume in the immediate future remains in the realm of contingency. Whitehead does believe, however, that the future must show some kind of internal determination. For example, we ordinarily assume that it will be spatio-temporally related to the present event, since this is what we mean by the future – that it is relative to something going on now.

III. SPECULATIVE PHILOSOPHY AND ITS METHOD

(i) *Speculative Philosophy*

Speculative philosophy[38] for Whitehead not only deals with a system of general ideas, but these ideas also require an empirical underpinning. Thus he defines speculative philosophy as "the endeavour to frame a coherent, logical, necessary system of general ideas in terms of which every element of our experience can be interpreted," so that "everything of which we are conscious, as enjoyed, perceived, willed, or thought, shall have the character of a particular instance of the general scheme."[39]

A philosophical system, then, needs to have a high degree of generality as well as a wide empirical application. It has also to be "coherent," and by this Whitehead means that "the fundamental ideas, in terms of which this scheme is developed, presuppose each other so that in isolation they are meaningless."[40] It is his contention that the meaning of an idea or thing is inextricably bound up with its relationship to other ideas and things. For example, when we proceed to explain the meaning of words it is always in terms of other, usually simpler words. Hence, for Whitehead not only are our ideas systematically related to each other, but so for that matter are the entities within experienced nature.

Whitehead enumerates some of the features which an adequate philosophical system ought to possess. It should be, he says, "coherent, logical, and, in respect to its interpretation, applicable and adequate."[41] In other words, its formal structure should be (1) coherent: every entity should be connected within the system; and as regards its interpretation, i.e., its empirical side, the scheme should be (2) applicable to some items of experience and (3) adequate: there should be no items incapable of being thus related as ele-

---

[38] Cf. *PR*, Part I, Chap. I, "Speculative Philosophy," pp. 3-23.
[39] *PR*, p. 3.
[40] *PR*, p. 3.
[41] *PR*, p. 3.

ments within the system – or as he puts it, "the texture of observed experience, as illustrating the philosophic scheme, is such that all related experience must exhibit the same texture."[42]

Whitehead does not believe that we can ever hope finally to formulate the metaphysical first principles.[43] He even doubts whether we can attribute metaphysical necessity to the most general principles of logic and mathematics, in view of the past mistakes regarding Euclidean geometry. Two reasons are given for our inability to arrive at a final formulation of metaphysical principles. These are (a) the weakness of our imaginative insight, which seems to be largely a conceptual matter; and (b) the deficiency of our ordinary language which lacks generality, since it has been designed to handle the ordinary things of life. Whitehead is here explicitly contrasting the concreteness of ordinary language with the generality of logic and mathematics.

As there is clearly a difference between the way we are aware of our everyday objects and the way we come to grasp metaphysical first principles, Whitehead remarks that we habitually observe the former by what he calls the method of difference: we are aware of tables, chairs, trees, etc., when they are present to our senses and we also take note of their absence. However, this method fails when we come to consider first principles, as unlike ordinary objects they are always present in our experience. Hence, in the discovery of first principles, we would not make much progress if we merely restricted ourselves to systematising detailed matters of fact in the Baconian manner. Owing to the generality of these principles we cannot be directly aware of them as we can of concrete objects. They can only be disengaged from the particular things around us as a result of a process of abstraction and generalisation.

Whitehead claims that a successful philosophical construction must start from the generalisation of certain features to be discovered in the various topics of human interest among which he includes not only physics, but also physiology, psychology, aesthetics, ethical beliefs, sociology and language.[44] He believe that one needs to formulate a scheme which should embrace every variety of experience, so that specific notions derived from some restricted group of facts should by generalisation be found capable of application to every other type of fact. The success of the imaginative generalisation (i.e. the scheme) is therefore to be tested by seeing whether it is applicable beyond the narrow range of experience from which it originated.

[42] *PR*, p. 4.
[43] *PR*, p. 5.
[44] *PR*, p. 5.

In *PR* Whitehead also held the view that mathematics employs the method of imaginative generalisation which he urges philosophy to adopt. In mathematical thinking, at least as Whitehead conceives it, we find (1) the generalisation of specific notions exemplified in particular instances, where the grasping of such notions might be compared to the seeing of a universal in a specific instance; (2) that in any particular branch of mathematics, the notions presuppose each other: for example, in geometry the propositions stand in implicatory relationships. And finally, the possibility always remains open that some such system may obtain an important application.[45] Time, he says, may be needed. Apollonius developed the theory of conic sections some eighteen hundred years before Kepler applied this theory to the planetary laws; and Reimannian geometry was developed before Einstein applied it to relativity theory.

From an early date Whitehead had been interested in the application (or interpretation) of abstract systems. This feature of his thought is best seen in *MC*, where he wishes to construct a number of alternative cosmologies. In *UA* Whitehead states a similar position: "The ideal of mathematics should be to erect a calculus to facilitate reasoning in connection with every province of thought, or of external experience in which the succession of thoughts or events can be definitely ascertained and precisely stated."[46] It might be argued that such a mathematical ideal is not one which a pure mathematician would share: it seems rather to typify the approach of a logician interested as, for example, Leibniz was, in the application of formal systems to particular subject-matters. Whitehead's attempt to construct a speculative philosophy – to frame a system in terms of which all our experiences can be interpreted, has overtones of the task he set himself in *UA* of formulating a calculus applicable to "every event phenomenal and intellectual, which can occur."[47]

(ii) *Other Philosophical Approachs*

As Whitehead sees it, the chief error of philosophy in the past has been what he calls overstatement.[48] Two forms of such overstatement are distinguished by him. The first is the fallacy of misplaced concreteness, which consists in considering experience in terms of such abstract notions as space, time, matter and mind. By assuming that these notions have an independent existence we give them a concreteness they do not possess, and thereby neglect the rest of

[45] *PR*, p. 6.
[46] *PR*, p. 7.
[47] *UA*, p. viii.
[48] *PR*, p. 9.

our concrete experience from which they have been abstracted. Whitehead believes that we can measure the success of a philosophy by its comparative avoidance of this fallacy,[49] which would seem to come to much the same thing as what has more recently been termed the category mistake.

The other form of overstatement "consists in a false estimate of logical procedure in respect to certainty, and in respect to premises."[50] An example of this is Descartes' method of dogmatically indicating premises and erecting upon these premises a deductive system of thought. Although this approach resembles Whitehead's own method of imaginative generalisation, Whitehead gives a very different status to the premises from which the system starts. He does not regard them as *a priori* in Descartes' sense, but as hypothetical. There are, he says, no precisely formulable axiomatic certainties from which we can start in philosophy. The verification of a rationalistic scheme is rather to be sought in its general success in practice, and not in the peculiar certainty of its first principles. Further, in metaphysics, unlike logic and mathematics, our chief interest lies not simply in the self-consistency of the system, but in finding some system whose structure is identical with that of our metaphysical situation.

Thus the ideal set for philosophy by Whitehead has some of the features of an axiom system, and it is clearly not achieved by the philosophy he develops in *PR*. Although some of the notions of his system have a high degree of generality, much of it appears to be descriptive. His actual system does not therefore always match up to the criteria which he claims a successful philosophical scheme should fulfil.

[49] *PR*, p. 10.
[50] *PR*, p. 10.

CHAPTER 9

PREHENSIONS AND SOCIETIES

THE THEORY OF PREHENSIONS

In this section we shall give some account of Whitehead's notion of a prehension,[1] which is a key concept in his later philosophy. He sometimes uses the term feeling as a synonym, which as thus used has a more general sense than it has in our ordinary use, where it specifically refers to our subjective affective states. Whitehead extends its meaning so as to describe both physical and psychological processes: simple physical feelings are also regarded by him as descriptive of energy changes in the physical world.

In support of his position Whitehead quotes Francis Bacon: "It is certain that all bodies whatsoever, though they have no sense, yet they have perception ... and whether the body be alterant or altered, evermore a perception precedeth operation; for else all bodies would be like one to another."[2] Bacon then distinguishes between *perception* or taking *account of* (which seems to be another way of saying that objects influence each other) and *sense* or *cognitive experience*. Whitehead remarks that the word *perceive* in our common usage is shot through with the notion of cognitive apprehension. Believing as he does, that we take account of things of which we do not have explicit cognition, he proposes to use the word prehension for such uncognitive apprehension.[3]

The concept of prehension (or physical feeling) which is elaborated in some detail in *PR* covers not only causal transactions in physical nature, but also the reactions of biological organisms to their environment as well as perception, cognition and judgment in man. Whitehead therefore finds it necessary to distinguish between simple physical feelings and complex ones. In the case of the former, we are concerned with the transmission of a form of

---

[1] Cf. *PR*, Part III, "The Theory of Prehensions," in which Whitehead develops the theory first sketched in *SMW*.

[2] *SMW*, pp. 55-56.

[3] *SMW*, p. 86.

energy from event to event in physical nature. Whitehead believes that physical events have in addition to their quantitative properties, certain affective characters – that they are "throbs of emotional energy."[4] Against the charge of anthropomorphism he would argue that physical science only deals with the measurable properties of such events and abstracts from their basic affective content. But he is careful to point out that the emotions experienced by us as human beings are not what he calls "bare emotion." They have already undergone a considerable amount of complex organisation and are accordingly termed by him "complex physical feelings." "But even so," he states, "the emotional appetitive elements in our conscious experience are those which most closely resemble the basic elements of all physical experience."[5]

Whitehead holds that it is therefore possible to draw an analogy between the transference of "sensory affective characters" from event to event in sense-experience (i.e. our awareness of a stream of sensory experience) and the transfer of energy in physical systems. He explains this further by saying that he is looking for elements in our direct experience in terms of which a systematic metaphysical cosmology can be constructed – elements not only found in our perceptual experience, but also in physical nature. As he succinctly puts it: "The key notion, from which such a construction should start, is that the energetic activity considered in physics is the emotional intensity entertained in life."[6] Not only then does this theory (or metaphysical system) apply to the basic elements in the physical world (simple physical feelings); it also applies to the elements of human experience or "complex physical feelings."

Whitehead describes a "simple physical feeling" as an act of causation.[7] By this he means that it can be regarded as a stream of influence analysable into the following components: (a) a past event or "the cause," which transmits its character to the present event; and (b) "the effect," that part of the stream which is the present event in the act of becoming the event which is future relative to it. Whitehead further says that "a simple physical feeling is the most primitive type of an act of perception, devoid of consciousness."[8] He would seem to be making here the same point as that made by Bacon above, namely, that the sensitivity of physical things to changes in their environment (i.e. the way one event causally influences another)

[4] *PR*, p. 163.
[5] *PR*, p. 228.
[6] *Nature and Life*, p. 96.
[7] *PR*, p. 334.
[8] *Ibid.*

is in some ways analogous to sense-perception in human beings. In its simplest form a positive prehension is such a transmission of influence in physical nature.

Whitehead has so far only been considering an artificially simple case of a "positive prehension," the linear transmission of a physical character from one event to another. But each event also stands in relationships (causal or otherwise) to the other events in nature, and is modified to however slight a degree by them.[9] The relationships which that event has to some other events in nature, and which in a sense it excludes, as well as the realm of alternative possibilities to which it is connected, are termed by him negative prehensions.

In his account of positive prehensions Whitehead is among other things concerned with the endurance of physical objects. He has defined physical endurance in *SMW* as "the process of continuously inheriting a certain identity of character transmitted throughout a historical route of events."[10] In other words, the character of each such event conforms to its predecessors. In *PR* Whitehead made the same point by saying that a simple physical feeling embodies (a) "the reproductive character of nature" and (b) "the objective immortality of the past."[11] It embodies (a) since it is this transmission of character (which may be sensory or physical) into the future which gives rise to the novel event, and (b) since it originates from the immediate past, from which the character has been inherited. It is this repetition of a self-identical pattern throughout a route of successive events, which for Whitehead gives physical objects their endurance, and makes us say that we are observing the same object over a period of time.

Whitehead tells us that our usual way of consciously prehending the world is by means of "transmuted feelings,"[12] namely, we are aware of it as made up of enduring objects characterised by definite sensory qualities. Our awareness of such objects is put under the head of "transmuted feelings," since he regards them as highly simplified versions of the physical activities going on in nature. Thus in our perception of the physical route of events constituting a stone "the immediate percept assumes the character of the quiet undifferentiated endurance of the material stone, perceived by means

[9] This doctrine is already to be found in his philosophy of nature. Thus he states, "the character of any event is modified (to however slight a degree) by any other electron, however separated by intervening events" (*PNK*, p. 98).

[10] *SMW*, p. 131.

[11] *PR*, cf. p. 336.

[12] Cf. *PR*, pp. 355-60, where he elaborates the "Category of Transmutation." This category covers the process whereby atomic entities are transformed into the continuous regions of the physical macroscopic world, as well as the integration of the neural and bodily processes in sense-perception into spatial regions illustrated by qualities.

of its quality of colour."[13] The approach to conscious perception then consists in the gain of a power of abstraction, so that the irrelevant multiplicity of physical detail is eliminated and emphasis is laid upon the systematic characters pervading events. In this way we become aware of a world of common sense objects as marked out by such qualities as colours, etc.

In *SMW* Whitehead states his theory of prehensions in more general terms. He tells us there that a "prehension" has a complex character, and that he thinks of it as a process of integration of the diverse aspects of other events into some particular pattern grasped into the unity of a perspective standpoint "here and now."[14] For Whitehead then an event is made up of (1) the pattern of aspects of other events integrated in its unity "here and now" and (2) the pattern of modifications it sets up in the neighbouring events. Thus, he goes on to say, "The ordinary scientific ideas of transmission and continuity are, relatively speaking, details concerning the empirically observed characters of these patterns throughout space and time."[15]

Whitehead further distinguishes in *PR* between (a) physical and (b) conceptual prehensions. In the case of sense-perception (a) would refer to the spatio-temporal aspects of other events integrated by the subject in his present perspective, and (b) to the aspects of their specific sensory qualities which characterise this perspective. Since each such experienced perspective is made up of these two kinds of prehensions, he terms it a "hybrid prehension." In the physical world, where we deal with simple physical prehensions, the notion of a physical prehension takes on the form of the transmission of energy from event to event, whilst "hybrid prehensions" refer to the origination and direction of energy, i.e., the way one type of energy is changed into another.

Whitehead's belief that our primitive emotional experience can act as a rough ontological guide to the physical activities or, to use Locke's term, "powers" in nature, is not one which would be acceptable either to those philosophers of science who define scientific entities as groups of instrument readings, or to those who look on such entities as "merely names for logical terms in conceptual formulae of calculation."[16] Whitehead has already made the point in *CN* that "the molecules and electrons of scientific theory are, as far as science has correctly formulated its laws, each of these factors to be found in nature."[17] This is still his view in his later philosophy, although he

13 *PR*, p. 107.
14 *SMW*, cf. p. 87.
15 *SMW*, p. 126.
16 *CN*, p. 45.
17 *CN*, p. 46.

would now say that not only have physical activities a structure, but also a primitive affective content.

II. THE SOCIAL THEORY OF REALITY

According to Whitehead reality exhibits a social order. By this he means that it is made up of "societies" or systems of events exhibiting regularities of pattern. Such societies manifest themselves in our experience as enduring objects. As he tells us, "an ordinary physical object which has temporal endurance is a society."[18] Whitehead terms the recognisable permanences in nature – tables, chairs, mountains, etc. – societies, since he regards each such object as a historical route of events pervaded by a self-identical pattern. Unlike a particular event, such as a flash of lightning which happens and passes, a society or an enduring object has a history expressing its changing relations to its environment. It is this endurance of a common pattern through a period of time which makes us say, when we observe the cherry tree in our garden, that we are observing the same tree today as we did yesterday.

Whitehead identifies the character of endurance possessed by an object with the social order of a nexus (or historical route). The self-identical pattern inherited throughout the individual members (or events) of that nexus is termed by him the defining characteristic of the society. A similar view is put forward in *CN*, where he points out that all you mean by stating that Cleopatra's Needle is situated on the Embankment, is that within the structure of events which forms the medium within which the daily life of Londoners is passed "you know ... that any chunk of this stream ... has the character of being the situation of Cleopatra's Needle."[19] This manner of defining the endurance of Cleopatra's Needle is identical with the definition of a society given in *PR*. The "continuous limited stream of events" corresponds to the "nexus", and what he terms a "chunk of the stream" to a "member of that nexus", i.e., an event. The permanent object, Cleopatra's Needle, may be identified with the common element of form illustrated in each of the members of that nexus. Whitehead's reason for calling such a self-identical object a society, is that he regards it as a set of related events bound together by a similar defining characteristic.

It follows from this, Whitehead says, "that whilst an object is in existence, it is always adding to itself with the creative advance into the future – a man

---

[18] *PR*, p. 47. The theory of societies is discussed in *PR*, pp. 134-153.
[19] *CN*, p. 167.

adds another day to his life, the earth another millenium to the period of its existence." "But," he goes on, "until the death of the man and the destruction of the earth, there is no completely determinate nexus which in an unqualified sense is either the man or the earth."[20] When we refer to the immediate objects of our perceptual existence, we usually refer to the realised society up to the present stage of its existence. Such objects may, of course, take on a very different look tomorrow: in the case of the cherry tree it may burst into blossom or be destroyed by lightning.

Whitehead distinguishes between those societies which have what he calls a "personal order," i.e., a simple self-identity, from those which have a complex structure, i.e., are made up of simpler objects or parts. An animal body made up of a multiplicity of cells would be an example of the latter; and as examples of the former, he gives (i) the life of a man and (ii) that of an electron. A man defined as an enduring percipient in terms of a stream of personal experiences or a self, and having a complex character in virtue of which he is considered to be the same enduring person from birth to death is then such a society. The other example of personal order, the life of an electron seems almost a matter of definition, since an electron would normally be assumed to be simple.

Whitehead next introduces the notion of "corpuscular societies." These refer to our everyday physical objects which are analysable into simpler objects such as molecules, electrons, etc. Each electron or molecule is itself to be thought of as a society (i.e. a pervaded route of events). In the case of an animal body we have groups of events patterned by electronic, molecular and cellular characteristics, which are co-ordinated within the wider pattern. Whitehead notes that as the individual characteristics of the included elements become more important in comparison with those of the whole pattern, so does that society become more "corpuscular," until the notion of a defining characteristic comes to cover the coordination of societies. A regiment, for example, may be regarded as more of a corpuscular society than say a man: since its defining characteristic refers rather to a peculiar kind of "social organisation." We can more readily take note of its component entities, the men who make up the regiment, than we can of the cells, molecules, etc., making up the individual men.

Corpuscular societies which have a complex tighter organisation are termed by Whitehead "structured societies." Material objects belong to the lowest grade of such societies, whilst animal and other living societies belong to the higher grades, since they are composed of living cells as well as inor-

---

[20] *AI*, p. 262.

ganic molecules. In considering the animal body, Whitehead points out that it provides the environment which sustains the sub-societies of cells, molecules, etc., as well as the stream of experiences which he calls the "regnant" nexus – since this nexus has some control over the functionings of the whole body. To take the case of our own body, it is to some extent dominated by our stream of personal experiences, desires, volitions and cognitions, which prevent us from being a mere slave to passing circumstances.

Whitehead outlines somewhat speculatively the way in which the "stream of experience" or "regnant" nexus is sustained by the animal body. A complex inorganic system of interaction, he tells us, is built up for the protection of the "entirely" living nexus," and the originative actions of the living elements are protective of the whole system. In its turn the reactions of the whole system provide the intimate environment required by the "entirely living nexus." On Whitehead's view then a living body consists of (i) inorganic systems of entities, electrons, molecules, etc., and (ii) a certain originality of character arising from the living elements which are in a sense sheltered by these systems.

Although a living nexus cannot sustain itself apart from its animal body, it may be so canalised, Whitehead argues, as to support a thread of personal order (or stream of conscious experience). Such a conscious personality, however, only seems to be characteristic of the higher animals. Whitehead expresses doubt whether we have grounds for conjecturing living personality in the lower forms of life. Nevertheless, he believes that although the lower animals and plants are not dominated by any such unique personality, the life arising from the individual cells within these societies is canalised into some faint form of social order. Thus in a tree, for example, where there is a lack of differentiation of function, we get something like a democratic control.

Whitehead refuses to accept an independent mind or soul presiding over the bodily events rather like an Uncle Sam presiding over, the destiny of the U.S. citizens. All the life of the body, he asserts, is the life of the individual cells. As there are millions upon millions of centres of life in each animal body, what needs explaining is the unifying control by means of which we have (a) unified behaviour and (b) consciousness of a unified experience.[21] Among the higher animals, he tells us, the bodily events are so co-ordinated that a stream of experience (defined in this general non-conscious almost Leibnizian manner) flows from the various parts of the body into the brain, where there is produced the presiding or conscious personality.[22] In

[21] *PR*, p. 151.
[22] As he puts it (*AI*, p. 271), "The whole body is organized, so that a general co-ordination of mentality is finally poured into the successive occasions of this personal society."

their turn the emergent stream of volitions, cognitions, etc., modify the events throughout the rest of the body and give rise to our unified behaviour.[23] As a consequence a man will behave differently from the physical objects around him. He can adapt himself to changing circumstances and obtain some control over his environment.

It might be argued that by assuming that the animal body is made up of societies of varying degrees of complexity, Whitehead overlooks that the defining characteristics of these societies are of different grades of generality. To talk as he does of a "subservient society" and a "regnant nexus" may perhaps have some significance when we consider the body from a purely biological point of view: one is then concerned with similar types of things. On the other hand, when he supposes that our direct experience and the cells and the molecules in our body are co-ordinated into the unity of a structured society, he does seem to be confusing entities of different grades of generality – physical and biological objects and our concrete experiences.

This objection does, however, involve a misunderstanding of Whitehead's position. He would say that to assume that the physical and biological objects postulated by science have an existence in themselves, is an example of his own fallacy of misplaced concreteness. It would be rather like saying that the grin remained behind after the Cheshire Cat had vanished. For Whitehead the cellular, molecular and experiential characteristics of our bodily events do not belong to radically different types – as do the matter and mind of Cartesianism. He believes that these different characteristics can be interrelated in the one society which is the animal body.

Nevertheless, there is some doubt as to whether Whitehead's distinction between the higher types of human and animal experience and the lower types said to occur in the physical world, is merely one of complexity. Whitehead describes the life occurring in the cells of the human body in terms of the appearance of novelty of character. In other words, the behaviour of the cell is different from what it would be if it were simply a summation of inorganic elements and not an organic whole. He would then seem to be referring here to the emergence of some new property (or properties). If this is so, then the higher types of experience are higher because they possess properties not reducible to those of the lower types.

---

[23] In *SMW*, (p. 98), Whitehead told us, "The concrete enduring entities are organisms so that the plan of the *whole* influences the characters of the various subordinate entities which enter into it." In the case of an animal, he went on, the mental states enter into the plan of the total organism, and thus modify the plans of the successive organisms down to the simplest organisms such as electrons.

CHAPTER 10

PERCEPTION AND BODILY DEPENDENCY

I. ATOMICITY AND CONTINUITY IN PERCEPTION

Whitehead is faced with the problem how, if according to the physicist the physical world consists of atomic activities, we nevertheless perceive it as made up of such common sense objects as trees, houses, tables, etc., which have a continuity about them. He notes, for example, the difference between the paving stone as perceived visually and its physical molecular activities. Pragmatically, he tells us, a paving stone is a hard, solid, static, irremovable fact. But this, he goes on, is a very superficial account if physical science is correct. Our sense-experience would then seem to omit any discrimination of the fundamental activities within physical nature.[1]

Whitehead assumes that in sense-perception this change occurs through a process of simplification (or transmutation), in which the characters illustrating the many individual events are fused into one sensory quality and integrated with a contemporary spatial region. He therefore believes that "Mentality is an agent of simplification,"[2] and that our perceptual experience is an immensely simplified version of physical reality. Nevertheless, he does recognise that the common sense world experienced by us is of immense value for our biological survival.

Whitehead's description of the way such physical activities are transmuted into the data given in conscious perception would seem to lean heavily on the findings of physiology. He tells us that "the bodily organization is such as finally to promote a wholesale transmutation of sensa inherited from antecedent bodily functionings, into characteristics of regions."[3] What

---

[1] *Nature and Life*, pp. 64-5.

[2] *AI*, p. 273. On the Category of Transmutation see *PR*, pp. 355-60.

[3] *AI*, p. 275. Owing to the generality of the Category of Transmutation it does not specify the actual process of simplification. In sense-perception Whitehead's account seems to be a paraphrase of the physiological findings. When describing the way the physical atomic activities are, as it were, "averaged out" into the continuous macroscopic objects of the physical world, Whitehead does lean heavily on contemporary physical theory.

Whitehead seems to be saying here is that the forms of energy, electromagnetic waves, sound waves, etc., impinging on our senses from the external world, are so modified that we finally become aware of them as clear-cut sensa illustrating definite regions of space. We should note that by the "inherited sensa" Whitehead is really referring to the physical stimuli which he assumes are the primitive counterparts of the perceived data.

Whitehead brings this point out further when he says that we must keep in mind the primary status of the sensa as qualifications of emotive tone. "They are primarily inherited as such qualifications and then by 'transmutation' are objectively perceived as qualifications of regions."[4] When Whitehead talks about sensa as qualifications of emotive tone in this context he is not referring to conscious perception, but rather to the physical stimuli, which during their transmission through the body become differentiated into the various types of sensory quality – sound, colour, "feel," etc., that mark out the contemporary spatial regions. During the bodily transmission these functions of sensa are gradually modified. To use his own words: "for the initial occasions within the animal body, they are qualifications of emotion – types of energy, in the language of physics: in their final functioning for the high grade experient occasion at the end of the route, they are qualities 'inherent' in a presented contemporary nexus."[5]

Whitehead would say that he was talking here about sense-reception: the changes produced in our bodies by physical events, and not about conscious sense-perception. And so, in his criticism of Locke's theory of perception, where sensory qualities are regarded as mental additions to physical nature, he can state: "But throughout the whole story, the sensa are participating in nature as much as anything else."[6] In the physical stage red, for example, is felt with the emotional enjoyment of its sheer redness, and as a result of the bodily transmission it becomes the clear-cut sensory quality directly observed by us. Or to take the case of smell: "The experience," he says, "starts as that smelly feeling, and is developed by mentality into the feeling of that smell."[7] Whitehead then seems to assume that there is something similar to these perceived sensa in nature, whether it be an "emotional enjoyment of sheer redness" or a "smelly feeling."

The view that physical events and sensory experience have a common structure is the basis of Whitehead's theory of symbolic reference. On this theory our sensory perspectives stand as symbols for the activities in the

[4] *AI*, pp. 314-15.
[5] *PR*, p. 446.
[6] *PR*, p. 462.
[7] *AI*, p. 315.

physical world. As he puts it, "symbolism from sense-presentation to physical bodies is the most natural and widespread of all symbolic modes."[8] Whitehead's whole notion of symbolic reference is then based on his belief that there are identical elements connecting human experience with the physical world. "This fact, that 'presentational immediacy' deals with the same datum as does 'causal efficacy,' gives the ultimate reason why there is a common 'ground' for 'symbolic reference.' "[9]

Although our sense-perceptions may symbolise the external world, there does not seem to be as much common ground between them as Whitehead assumes. Whitehead would, of course, also argue that we have direct awareness of the physical events in nature through perception in "causal efficacy." However, he is still left with the problem of coordinating these two systems of perception. The physical account in itself does not explain why a physical stimulus should give rise to a sound, colour or smell. With a suitable apparatus or specially selected sense-organs it might give rise to any one of these. The "sensa" as they manifest themselves in physical nature can give no indication as to what kind of sensory qualities they may be transformed into. Whitehead would probably say that though the "physical sensa" are to be regarded as forms of emotional intensity, they are relatively undifferentiated in comparison with the observed sense-qualities. The differentiation which does occur is due to the sense-organs of the human organism.

The attempt to use the notion of structural correspondence between physical stimuli and sense-qualities as a basis for symbolic reference does run into certain difficulties. Russell, who advocated a somewhat similar approach but not Whitehead's primitive qualitative correspondence, stated, "As to intrinsic character, we do not know enough about it in the physical world to have a right to say that it is very different from that of percepts; whilst as to structure we have reason to hold that it is similar in the stimulus and the percept."[10] However, if in dealing with the physical world the scientist is, as Russell has so often said, only concerned with structure, the question of intrinsic character is quite irrelevant; and furthermore, there seems no *a priori* reason to believe "that the stimulus must possess whatever structure is possessed by the percept." The situation is different in the case of Whitehead, since he argues that there is such a correspondence of intrinsic characters. However, he would not locate the perceptual pole of this correspondence in the sensory data as Russell does, but in our perceptions of causal efficacy – in our experience of the "powers" within nature.

[8] *Symbolism its Meaning and Effect*, p. 5.
[9] *PR*, p. 243.
[10] Bertrand Russell, *Analysis of Matter*, p. 400.

II. BODILY DEPENDENCY AND SENSORY PERSPECTIVES

In his philosophy of nature writings Whitehead had asserted that we are aware of a common external nature which also contains within it such qualities as colours, sounds, smells, etc. But when he analyses the immediately given data of perception, it turns out that these qualities have a bodily dependency. In the relation of ingression there is always an implicit reference to a percipient event (or bodily event) which is happening simultaneously with the event in which the sense-quality, e.g., the colour red is situated.[11]

In his philosophy of nature Whitehead distinguishes between the way we apprehend events and the way we recognise objects, but he tends to put both kinds of awareness under the heading of sense-perception. In *PR* he brings in the two perceptual modes of "presentational immediacy" and "causal efficacy" to cover these two different kinds of perception. The difference between the sensory perspective (or present locus) and the duration (or field of physical activities) is thus made more precise. We usually assume that our sensory experience is closely related to the actual physical events now happening, and tend therefore to identify them. But Whitehead points out they are quite distinct. The observed perspective largely depends upon our bodily functioning, which though usually excited by the antecedent physical events, may yet, as in the case of hallucinations, arise through abnormal conditioning events such as the use of drugs.

Whitehead now discusses in more detail the important part played by our bodily experience in perception. When we are aware of some external region as illustrated by specific sense-qualities, we have also direct awareness of our sense-organs as efficacious in perception. Thus when we see a man in front of us, we also have the vague awareness that we see him by means of our eyes. Most philosophical accounts of perception, Whitehead argues, delete the bodily functioning and concentrate on the immediately given coloured patches in our perceptual experience.

He notes that the bodily parts experienced as causal in perception are perceived with greater distinctness by means of bodily sense-data, for instance, the slight eye-strain involved in seeing or the kinaesthetic sensations in our arm when we push against obstacles. But we need to distinguish between the arm as an experienced unit bounded by certain visual data of colour and shape, and the kinaesthetic data occurring within it: the latter are still for Whitehead perceptions in the mode of presentational immediacy. On the other hand, when we move an arm, the more deep-rooted awareness

---

[11] *PNK*, cf. pp. 84-85.

of it as efficacious in action belongs to the sphere of causal efficacy.

And furthermore not only do we experience causal efficacy in bodily motor activity, but Whitehead would also claim that we have direct awareness of a surrounding efficacious world. As, he puts it, "So far as concerns the causal efficacy of the world external to the human body, there is the most insistent perception of a circumbient efficacious world of beings."[12] Whitehead is not then simply putting forward an activity theory of causation as Maine de Biran did, who believed that our feeling of volitional effort during our performance of bodily tasks, gave us an awareness of the causal order in nature. Whitehead argues that we do have a direct awareness of this causal order, although as experienced by us it is extremely vague. "The definite discrimination, which in fact we do make, arises almost wholly by means of symbolic reference from presentational immediacy."[13]

The function of the perceived sense-data, as Whitehead sees it, is to connect the immediately past events (which were causal for them) with the events now happening in nature, so that they come to act as signs for them. In this way the sensory data have some relevance to the contemporary happenings in nature. For example, we are driving a car and as we approach the traffic lights they turn to red: the red light warns us of the danger ahead if we carry on, and we put our foot on the brake and stop. On the other hand, the red spot of light which we see when we look up at the night sky, may not have this immediate relevance, as it may be due to a star which went out of existence many millions of years ago.

For Whitehead, then, sensory data are related to the functionings of our bodies as well as to the antecedent external events. He claims that during the perceptual process our sensory perspective is mapped out in terms of spatial regions characterised by sense-qualities. The changing event which is our body is given a position "here-now," and is marked out by sensory qualities of colour and shape as well as exhibiting an intimate feeling of bodily awareness. Further, the region where the perceived objects are said to be situated (the focal region), is given a position "out there." The peripheral spatial regions extending beyond the observer and the perceived objects are mapped out in a similar fashion. What Whitehead is describing here albeit very schematically, is the mechanism which he believes underlies our perception of the external world and our perception of bodily efficacy[14] – a mechanism which he would no doubt admit was largely hypothetical.

Whitehead's view that our perceptions are perspectives of some real

[12] *Symbolism its Meaning and Effect*, p. 65.
[13] *Ibid.*, p. 65.
[14] *PR*, Part IV, Chap. IV, "Strains," pp. 439-456.

objective world is most clearly stated in *SMW*. Considering our perception of some green leaves he points out that, "green is not simply at *A* where it is being perceived, nor is it simply at *B* where it is perceived as located; but is present at *A* with the mode of location in *B*."[15] If you wish to have an example of this, he goes on, you merely need to look into a mirror and to see the image in it of some green leaves behind your back: green is then located behind the mirror, but if you turn round and look at the leaf "green has now the mode of being located in the actual leaf."

The fact that in our perceptions we see beyond ourselves and away from our particular perceptual standpoint "here and now," is summed up in Whitehead's almost Leibnizian statement that every spatio-temporal standpoint mirrors the universe. He points out that if one thinks of it in terms of our naive experience "it is a mere transcript of the obvious facts, you are in a certain place perceiving things. Your perception takes place where you are, and is entirely dependent on how your body is functioning."[16] If, then, this perception conveys knowledge of the world stretching beyond us it must be, he concludes, because our bodily life unifies in itself aspects of the universe.

However, one may well be puzzled by the connection supposed to exist between say our observation of the landscape stretching away in the distance and the light travelling to one's retinae from the objects in our physical world. Whitehead would say that the immediately perceived perspective, though observed as stretching beyond us, is yet bodily related: that there is a continuity between the body, conscious experience and nature. In postulating such continuity Whitehead clearly assigns to our bodily functioning a major role in perception. But on the other hand, he also states that from the point of view of direct experience, the physiological account of perception is a "tissue of irrelevancies." Even if Whitehead postulates, as he does, a special mode of perception (i.e. causal efficacy) which give us an awareness of our bodily functioning, there is still a considerable difference between our experience of bodily functioning and the physiological account.

In discussions on perception, there would seem to be two senses in which the term "body" may be used, and their blurring may lead to confusion. There is (1) the notion of the body as we immediately experience it through our kinaesthetic sensations and more deep-seated organic feelings; this might be termed the "lived body." And there is (2) the notion of the body as a biological entity as it enters into physiological and organic descriptions – in other words, the scientific body. The perceived data on Whitehead's

---

[15] *SMW*, p. 88.
[16] *SMW*, pp. 111-112.

account are certainly beyond the body in sense (1) but not in sense (2). If we criticise him for not adequately distinguishing these two senses, he would no doubt reply that the physiological account is concerned with certain abstract aspects of our concrete bodily functioning – of which we are aware through our experience of the lived body.

CHAPTER 11

PROPOSITIONS AND JUDGMENTS

I. PROPOSITIONS AND EXPERIENCE

As we have seen, Whitehead believes that in sense-perception we only obtain a simplified version of physical reality. We experience our common-sense world in what Whitehead calls perception in the mode of presentational immediacy, as made up of spatial regions marked out by clear-cut sensory qualities – coloured patches, tactual expanses, volumes of sounds, etc. – occurring at definite moments of time. On a somewhat different level, images occur in such mental states as memory and imagination. Although images have a similar structure to our percepts, i.e., they are qualities predicated of certain places and times, Whitehead would not accept Hume's view that they are merely faint copies of these percepts. He does allow that we have a capacity to generate new types of ideas.

Whitehead uses the term "proposition" as a collective noun to describe the data given in our perceptions and mental states.[1] This may seem a very odd use of the term, and one at variance with its normal logical usage where it refers to indicative statements. Whitehead would justify his usage by saying that the notion of a spatial region illustrated by qualities is at the root of the traditional substance-quality concept. He makes this point when he analyses a proposition in terms of (a) the "logical subject" – a spatial region occurring at a particular time, and (b) its "predicative pattern" – the eternal object or quality which is predicated of that logical subject. A proposition in this Whiteheadian sense is then a translation of the Aristotelian subject-predicate form of proposition from a logical to an epistemological level.[2]

---

[1] For an account of Whitehead's "Theory of Propositions," see *PR*, Part II, Chap. IX, pp. 260-94, "The Propositions," and Part III, Chap. IV, "Propositions and Feelings," pp. 362-375.

[2] A similar doctrine to that of propositions appears in *PNK* (Chap. XXII, "Figures," pp. 190-194), under the head of sense-figures, namely, a spatial region in which sense-

Further, such propositions in their perceptual occurrence, are said by Whitehead to act as symbols for the physical activities in nature. Whitehead believes that not only do we think symbolically, but that we also perceive in terms of symbols. The coloured shapes, etc., observed by us act as symbols for the primitive feeling elements in our experience, through which we obtain awareness of the causal efficacy of nature. Thus, he tells us, when we see a "grey stone" the grey refers to the grey shape immediately before our eyes, but the word stone has reference to other elements in our experience – the feeling of the efficacy of the physical stone in nature.

Our actual perception of a proposition involves what Whitehead calls an "affirmation-negation contrast,"[3] which is another name for the symbolic reference just noted. The negative side of this contrast is the proposition itself – a pattern of qualities marking out a specific spatial region at a particular time. As a proposition can be entertained in imagination or thought as a pure possibility, it can be said to be a negation since it is then excluded from any actual perceptual exemplification. Whitehead takes the affirmative side of this contrast to be the efficacious background of bodily feeling, through which we obtain an awareness of the physical activities in nature. In his earlier philosophy Whitehead spoke of our immediate awareness of the passage and activity of events in nature and noted that knowledge of the objects which mark them out, is obtained through the intellectuality of sense-recognition.[4] There then seems to be a certain resemblance between these two aspects of perception – the dynamic and conceptual – and the "affirmation-negation contrast" (or conscious perception) of *PR*.

If the function of propositions in perception is to serve as symbols for the physical facts in nature, then they may either act as faithful guides to such facts, or they may mislead us and give rise to illusion. On Whitehead's view, as we have seen, there is a close connection between our conscious perceptions and events in the physical world. This comes out in his enumeration of the factors necessary for our experience of a proposition.[5] They are: (1) a physical feeling (i.e. an external physical event) from which is derived (a) the "logical subject" or specific spatial region at a definite moment of time

objects are located at moments of time. Whitehead notes that sense-figures possess a higher perceptive insistency than the corresponding sense-objects. "We first notice a dark-blue figure and pass to the dark-blueness" (*PNK*, p. 192). In our perception of a sense-figure, say a dark-blue surface, we deal with what has been termed a particular. It is only in thought that we pass to the dark-blueness which is the universal.

[3] *PR*, cf. p. 377.
[4] *CN*, cf. p. 189.
[5] *PR*, cf. pp. 368-70.

and (2) a physical feeling from which (b) the eternal object or "predicative pattern" is obtained.[6] Although in normal perception (a) and (b) are usually derivative from the same (or very similar) events, this is not the case in abnormal perceptions, e.g., hallucinations. In such cases the predicated qualities are originated by the subject himself,[7] whilst the spatial regions they illustrate, are directly relevant to the events now happening in nature. It is said that after having taken mescalin Jean-Paul Sartre imagined that he was being trailed through the streets of Paris by a lobster. The hallucinatory lobster would have been specific to Sartre's own experience, whilst the places and times at which he saw it, would be closely related to the actual events now happening.

But even in everyday normal perceptions some of the qualities attributed to the perceived objects are contributed by the observer himself. In Whitehead's earlier philosophy, such qualities played an important part in normal perception: thus e.g., we may see a piece of velvet and imagine we are feeling it. Whitehead would seem to be making here the old distinction between sensation and perception: that in perception we complete the fragmentary sense-data by such subconscious sense-presentations, so that we perceive meaningful perceptual objects. Thus although we may only see the front part of an object, say a chair, we assume that it has a back, which we are not directly observing at that moment, that it is hard to the touch, that it can be sat on, etc., and we adapt our behaviour to it accordingly.[8]

Whitehead distinguishes a number of different kinds of propositional feelings (i.e. our experience of propositions). The two main types are: (i) perceptive propositional feelings which cover both authentic (veridical) perceptions and unauthentic (illusory) perceptions, and (ii) imaginative propositional feelings – the images occurring in such mental states as memory and imagination. This division might seem to resemble Hume's classification of the contents of the human mind into impressions and ideas. But on Whitehead's view, Hume's impressions are already highly complex. A simple Humean impression of sensation such as red would be regarded by Whitehead as an abstraction from our massive total experience.

In the case of (ii) imaginative feelings, although the subject originates the imaged data himself, there is usually a vague awareness of the contem-

---

[6] The eternal objects are derived according to the conditions laid down by Categoreal Obligation IV, Conceptual Valuation, *PR*. cf. pp. 350-352.

[7] These "reverted" eternal objects are derived according to the conditions laid down by Categoreal Obligation V, Conceptual Reversion, cf. pp. 352-353 and p. 368. They resemble the conveyed sense-objects of *CN*, cf. pp. 154-156.

[8] *CN*, cf. p. 155.

porary events which act as a reference point for these data. Such images as occur, for example, in day-dreams do not usually have a very close relevance to the times and places about which they are predicated. As Whitehead remarks when speaking of the imaginative use of propositions: it is more important for a proposition to be interesting than true. When e.g., we entertain propositions during our reading of imaginative literature they are "tales that might be told about actual entities."

To illustrate the different ways in which we may be aware of the same proposition, Whitehead takes the statement "Caeser crossed the Rubicon."[9] He first considers one of Caeser's soldiers actually observing Caesar's crossing, and able therefore to verify directly this proposition. Whitehead next considers the same soldier who at a later date revisits the river and is reminded of Caeser's crossing. The soldier now entertains the proposition imaginatively, and locates relatively to himself the past occurrence. His attitude towards the proposition will be, however, radically different from his earlier one. It is perhaps now experienced with a strong feeling of belief, whereas on the earlier occasion his attitude was one of simple acceptance.

Unauthentic perceptive feelings[10] for Whitehead cover a range of experiences, extending from simple illusions such as mirror images to wild hallucinations; they hence tend to shade into imaginative feelings. In imagination we have a certain freedom in selecting the kinds of ideas we entertain. When writing a piece of science fiction, we are free to imagine what the world was like millions of years ago, or what the state of our civilisation will be in A.D. 3000. On the other hand, unauthentic feelings are due to what Whitehead calls a "tied imagination": the data usually have some relevance to the events of which they are predicated, although the subject may have little control over their appearance. In seeing, for example, the image of a chair in a mirror, we see it in front of us (although it is really behind us) and we may even try to sit on it until we realise that we are looking into a mirror. In more extreme cases of hallucination, the data usually result from abnormal conditions in the subject, and may have little direct relevance to the events of which they are predicated, as when a chronic alcoholic sees pink rats in his room during an alcoholic bout.

Whitehead's position as far as the nature of authentic and unauthentic perceptive feelings is concerned, has much in common with his treatment of "normal" and "delusive" perceptions in his nature philosophy.[11] He there distinguishes between "active" and "passive" conditioning events.

---

[9] *PR*, pp. 276-7.
[10] *PR*, cf. pp. 381-2.
[11] *PNK*, 23.3. 24.6; *CN*, pp. 151-155.

Thus in our perception of a chair, the active conditioning events are those belonging to the chair in the physical world. But if, on the other hand, we observe the image of the chair in a mirror, the situation where it is observed may be said to be a passive conditioning event. In hallucinatory perceptions, the active conditioning events are usually to be found in the observer.

II. BELIEF AND JUDGMENT

Whitehead next discusses what he terms "intellectual feelings," or conscious experience. He does not, however, believe that conscious experience has the importance that some philosophers have assigned to it. "Consciousness," he says, "flickers; and even at its brightest, there is a small focal region of clear illumination, and a large penumbral region of experience which tells of intense experience in dim apprehension."[12] Intellectual feelings are classified by Whitehead under two heads, (a) conscious perceptions and (b) intuitive judgments.[13] These refer to the "perceptive" and "imaginative" feelings of propositions when integrated with our bodily experience in the form of a contrast. What he is essentially concerned with here are the impressions and ideas of David Hume, except that Hume regarded impressions as simple and unanalysable. But as we have seen for Whitehead they are related to certain antecedent physical occurrences, and in our experience always occur against a background of bodily feeling. Even in our most theoretical speculations we cannot completely abstract ourselves from our bodies.

Whitehead believes that imaginative activity plays an important role in our experience. For example, as we have already seen, perceptual objects have usually associated with them certain subconscious (or conveyed) sense-qualities. This feature of our experience is much more evident in delusive perceptions, where there is usually a set of "conveyed" sense-qualities, which may have little direct relevance to the events they illustrate. In such cases, he says, conscious perceptions tend to approximate to intuitive judgments;[14] they become imaginative notions about things. An example of this would be Sartre's hallucinatory lobster. Intuitive judgments also cover cases of memory; in memory the "recollected qualities" resemble those of past events. The subject's attitude to them has what Whitehead calls a belief character,[15] a belief that the recollected qualitative pattern is similar to that

[12] *PR*, p. 378.
[13] Cf. PR, p. 376 and pp. 382-9.
[14] *PR*, cf. p. 384.
[15] *PR*, cf. p. 378.

observed in the past. And it is this close relevance to the character of pre-
vious occasions which gives memory its characteristic fixity and order.

Whitehead next discusses "conscious imagination," namely, intuitive
judgments whose emotive pattern (i.e. subjective attitude towards the experienced datum) is dominated by indifference to truth or falsehood. In
these cases the imagined data have still some relevance to our actual world.
"We are," he says, "feeling the actual world with the conscious imputation
of imagined predicates, be they true or false."[16] Although these imagined
predicates have this general reference to the world, they may yet have very
little direct relevance to specific events in the past or the present. C.P. Snow
makes a somewhat similar point, when he refers to the setting in which the
action of his novel *The Affair* takes place. "This fictional college stands upon
an existing site, and its topography is similar to that of an existing college,
though some of the details are different. That is the end, however, of my
reference to a real institution."[17]

If we compare imagination to reasoning or remembering where there is
attention to truth, we see that in imagination the subject's attitude towards
the imagined data is usually one of indifference to truth.[18] When we engage
in imaginative activity as when reading a work of fiction, there is a readiness
to overlook the facts of our mundane situation. In imagination we are not
essentially concerned with the truth or falsity of the imagined data. Unlike
memory, there is usually no adherence to the characteristics of some past
event: there is the fluidity of imagination as contrasted with the order and
pattern of memory.

Whitehead goes on to examine intuitive judgments in which there is attention to the truth – where we take account of the truth character of the proposition. He distinguishes three kinds: (1) affirmative (*yes-form*), (2) negative (*no-form*) and (3) suspended (*suspense-form*). The form taken by each
type of judgment is dependent upon the different psychological attitudes
which occur in the subject, when he compares the entertained proposition
with the experienced facts.[19]

In (1) affirmative intuitive judgments, the "imagined pattern" is seen to be
identical with that of the events to which they refer,[20] as in memory, where we
may feel that the complex image we entertain resembles some earlier occurrence. Our experience will then contain an attitude of belief in the proposi-

---

[16] *PR*, cf. p. 388.
[17] C.P. Snow, *The Affair*, Macmillan, 1960, p. 2, Note.
[18] *PR*, cf. p. 389.
[19] *PR*, cf. pp. 382-3 and p. 385.
[20] *PR*, cf. pp. 382.

tion. For example, when we believe that the Second World War began in 1939, or that our next door neighbour's Alsatian dog is now deceased. On the other hand, in (2) negative intuitive judgments,[21] the imagined pattern conflicts with the facts; this shows itself in an attitude of disbelief in the truth of the proposition.[22] Like Mother Hubbard's dog we may disbelieve that there is food in the cupboard when we open it and find it empty. Whitehead notes that these two cases of intuitive judgment – the affirmative and negative judgments – together with conscious perception correspond to what Locke calls knowledge.[23]

The more usual form of intuitive judgments are (3) suspended judgments.[24] In these cases the imagined pattern is neither compatible nor incompatible with that of the perceived fact. Our attitude to it will therefore involve a certain mental reservation as to the truth of the proposition. As Whitehead says: "Our whole progress in scientific theory, and even in subtility of direct observation, depends on the use of suspended judgments."[25] The strength of our belief in the theories we hold about natural and social phenomena as well as our conjectures about the common sense world, is dependent upon the degree of compatibility between these theories and conjectures and the facts with which they are compared.

Whitehead tells us that his theory of judgment can equally well be described as a correspondence theory or a coherence theory,[26] and this is connected with the distinction he draws between a judgment and a proposition. He considers a judgment to be a purely psychological attitude in the judging subject. A proposition, however, has an objectivity about it: it is not restricted to that subject, for it can, he says, "constitute the content of diverse judgments by the diverse judging entities."[27] We can, for example, become aware, and so for that matter can others, of the truth of the proposition that Watson and Crick discovered the helical structure of chromosomes, by remembering that such an event actually occurred or by consulting an appropriate reference work.

Whitehead's claim that his theory of judgment can be taken as a correspondence theory stems from his description of judgment as the complex feeling (having the character of belief or disbelief) arising from our aware-

[21] *PR*, cf. p. 382 and p. 387.
[22] *PR*, cf. p. 385.
[23] *PR*, cf. p. 387.
[24] *PR*, pp. 382-3.
[25] *PR*, p. 388.
[26] *PR*, cf. p. 269.
[27] *PR*, p. 273.

ness of the compatibility or incompatibility of the proposition and the objective (or fact).[28] In an affirmative judgment we are then aware of a truth relation existing between these two components, and one which can be considered by other judging subjects. Most of us can, for example, entertain the proposition "Napoleon was defeated at Waterloo" and compare it with our knowledge of the historical facts. But as such a judgment is concerned with the conformity of two components within one experience, it may also be regarded, Whitehead says, as illustrating a coherence theory of truth.[29] "A judgment," he tells us, "is a synthetic feeling, embracing two subordinate feelings in one unity of feeling."[30]

A proposition for Whitehead can then be true or false and a judgment can be correct, incorrect or suspended. In the case of the proposition we are only concerned with its truth relation to the objective which makes it true or false. In a judgment, however, we deal with the subject's experience of the judgment's correctness or incorrectness. In judging the truth of a proposition, we are aware of the relation of identity which exists between the proposition and the objective, and which shows itself in our experience as a feeling of harmony or correctness. On the other hand, when we judge it to be false, our awareness of the diversity of the components exhibits itself as a feeling of discord or incorrectness.

In intuitive judgments we verify propositions by reference to empirical facts. There are, however, other judgments, particularly in logic and mathematics, where attention is directed rather to the proposition and its relations to other propositions, and which Whitehead puts under the head of derivative judgments.[31] As the truth of these propositions are independent of empirical verification, logical criteria are used to prevent error from creeping into such judgments. "Logic," Whitehead says, "is the analysis of the relationships between propositions in virtue of which derivative judgments will not introduce errors, other than those already attaching to the judgments on the premises."[32]

In derivative judgments then Whitehead is primarily concerned with deductive reasoning, where the validity of each step in the process is dependent upon logical criteria. In such judgments the attitude of assent or dissent will therefore depend upon whether or not the derived propositions stand up to such logical tests. Derivative judgments cover the whole range of our reason-

[28] *PR*, p. 269.
[29] *PR*, p. 270.
[30] *PR*, p. 273.
[31] On the difference between intuitive judgments and derivative judgments, cf. *PR*, p. 271.
[32] *PR*, p. 272.

ing from our day-to-day verbal arguments to the more complex chains of logical and mathematical reasoning, where the propositions have only a structural form and are abstracted from qualitative content.

CHAPTER 12

CAUSATION AND PERCEPTION

I. WHITEHEAD'S CRITIQUE OF HUME

More ink has probably been spilt in philosophy over the concept of causality than any other single concept. For Hume, for example, the causal relationship was a habit of expectancy, for Kant a conceptual category and for Maine de Biran a feeling of subjective effort. In recent years under the influence of phenomenology some writers, in particular the Gestalt psychologists and the Belgian psychologist Michotte,[1] have argued that we have a direct perception of causality much in the same way as we perceive shape and movement. Michotte gives the example of a knife cutting a slice of bread. We do not, he says, just see two independent movements, the advance of the knife and the cutting of the bread; we have a specific causal impression of the two movements as essentially and temporally coordinated, forming a continuous process in which one is productive of changes in the other.

Whitehead too believes that there is such a phenomenon as perceptual causality, and his views on this question are very similar to those of Michotte and his school. For all practical purposes, he tells us, we never doubt that the event now happening conforms to those in its immediate past. We directly observe a process of development much as we do particular colours and sounds, but it is a characteristic belonging to a whole situation rather than to any sense-quality. We would be unable to adapt ourselves to our immediate environment if we could not make short-term forecasts. Thus on observing the earlier part of such a process we can anticipate its later phase and adapt our behaviour accordingly: for example, when driving we brake when the car in front slows down.

The problems which have arisen in philosophy as to the relation of causation and perception, arise, Whitehead believes, from an inadequate account

[1] A. Michotte, *The Perception of Causality*, translated by T.R. & Elaine Miles, London, Methuen, 1963.

of our immediate experience by such philosophers as Hume and Kant. Their denial of any awareness of direct causal determination stems from their deletion of the perceptual process of passage from the experienced temporal process.[2] If the only data given to us in perception are, as they maintain, a succession of sensa, it becomes necessary to explain why nevertheless, we are persistently aware of a passage of events causally conforming to each other. To overcome this difficulty Hume tried to reduce our experience of causality to a psychological habit and Kant assumed that it was due to a conceptual category imposed upon our experience.

Whitehead is particularly concerned with examining Hume's account of causality.[3] According to Hume, he says, the appearance of certain antecedent percepts, for example, those constituting a red hot stove, leads us to expect the consequent with which they have in the past been associated – in this case a feeling of pain when we put our hand on the stove. On Hume's view causation is then simply a habit due to the frequent association of past impressions. It is we who read causation into our impressions by projecting on to them a feeling of expectancy. Causality then becomes a peculiar fact about the way our minds work rather than about perceived nature.

By way of testing the truth of Hume's theory, Whitehead considers the case of reflex action. As a particular example of such a reflex he takes the blinking of a man's eye when a light is suddenly shone in the dark.[4] Whitehead asks us to examine closely the man's subjective experience. Restricting ourselves to the sequence of sensa as Hume does, they are: flash of light, feeling of eye closure, instant of darkness. But, Whitehead goes on, if we question the man he will also assert that he distinctly felt the flash give rise to the blink. In other words, he has a direct awareness of causal determination – that one experience has produced the other: it is because of this that he can distinguish that the flash came before the blink. The whole occurrence flash-making-one-blink, hangs together in a connected pattern.

Whitehead's position as regards perceptual causality, as we have already seen, is one he shares with the Gestalt school of psychology. W. Köhler[5] has, for example, stated that on Hume's view feeling uneasy near a hot radiator would be one experience, and then as an entirely separate experience we might feel ourselves moving away from the radiator. Köhler considers this analysis of causal perception to be highly superficial, as these two experiences are not independent of each other: they belong together in one natural con-

---

[2] *Symbolism its Meaning and Effect*, cf. p. 41.
[3] *PR*, cf. p. 245.
[4] *Ibid.*, cf. pp. 245-6.
[5] *Gestalt Psychology*, New York 1929, cf. p. 384.

text. We need to make a definite analytic effort to think of them as isolated experiences.

However, we still have to deal with Hume's denial that there is such a direct perception of causal determination.[6] In the case of the man blinking when a light is shone in the dark, Whitehead argues that Hume's reason for this denial is based on his *a priori* assumption that we are only aware of a succession of sensa. As a result he attempts to reduce the man's feeling of compulsion to one of expectancy that the blink will follow the flash, namely, to the feeling of a habit. But how, Whitehead asks, can we feel a habit when we cannot experience a cause; on Hume's view are they not equally metaphysical chimeras? Despite the superficial cogency of Hume's argument, he confuses a habit with a perceived feeling of it.

Further, Whitehead would argue that causal determination is not a specific sensation but a characteristic belonging to a whole occurrence extending over a period of time. It is only one of the many qualities given in our perceptual experience which do not fit into the sensationalist account, an account that postulates atomic sensations bound together by association. As the Gestalt school has pointed out, our experience contains many such "holistic" qualities: for example, a circle is "round," a decorative pattern is "symmetrical," etc. To quote Sartre, "Being dreadful is a *property* of this Japanese mask ... and not the sum of our subjective reactions to a piece of sculptured wood."[7]

In the case of perceived causality Whitehead would say that he is describing what the naive individual actually observes – that he is directly aware of such causal determination. He might also go on to say that he was merely indicating the crude perceptual data from which the more refined causal notions of science are derived, just as he has indicated those elements in our experience from which we obtain the physical notions of space and time. He therefore believes that this awareness of direct determination – of the way one perceptual situation grows out of another – is some guide to the causal processes in the physical world.

Whitehead would agree that in a certain sense we do not see *A* cause *B*, if by "cause" is meant the kind of cause postulated by the scientist, or a logical relation of material implication. He would take the former notion at least, as a limiting conception obtained from our crude perceptual data. Causation as thus used in science and perhaps by some philosophers who

---

[6] *PR.* cf. pp. 246-247.

[7] "Intentionality: A Fundamental Idea of Husserl's Phenomenology," translated by Joseph P. Fell, *Journal of the British Society for Phenomenology*, Vol. 1 No. 2, p. 5.

seem to seek the pure logic of it, is very different from perceptual causality as described by Whitehead.

II. CONTEMPORARY STUDIES IN PERCEPTUAL CAUSALITY

As Whitehead's discussion of causality is based on what he takes to be descriptions of our immediate experience, we need to examine more closely the nature of perceptual causality. This may best be done by referring to Michotte's work. Michotte's great merit is that he has given an experimental analysis of perceptual causality. He found that causal impressions which he could produce under laboratory conditions, could be classified into two basic types: "launching" and "entraining." In the launching experiment a black object *A* moves towards a red object *B*, then starts and moves away at a somewhat slower speed and then stops. The observer clearly sees object *A* bump into object *B* and send it off or launch it. In the entraining experiment, as soon as object *A* touches *B*, *B* in its turn starts to move off at the same speed. There is the definite impression that object *A* carries object *B* along or entrains it.

These causal impressions remained unaffected, even when the same experiment was tried out hundreds of times on the same subjects. When certain conditions of speed, position and time-interval were satisfied his subjects received an impression of causal interaction between the two movements. For example, the movements of objects *A* and *B* have to occur in quick succession and a change in *A* has to be followed by a change in *B*. As against Hume, Michotte shows that habit and expectancy are not the crucial factors in giving rise to a causal impression. If the stimulus conditions are right, the causal impression occurs right away and otherwise it will not occur, however often the experiment is repeated. Hence, Whitehead's contention that the notion of a habit is ineffective in explaining why we do observe causal determination, seems to be borne out by the empirical facts adduced by Michotte.

Michotte assumes, like Whitehead and the Gestalt psychologists, that in some way phenomenal causality is ontologically prior to our understanding of physical causality, although there is a correspondence between them. As he remarks "It is of course from phenomenal data that the physicists' 'world' is constructed."[8] Michotte has told us elsewhere that there seems to exist in the phenomenal world, a sort of prefiguration of such abstract no-

---

[8] *The Perception of Causality*, p. 307.

tions as substance, reality, causality, etc., in terms of which the unsophist-
icated person spontaneously adapts himself to the world.[9]

In an essay "Michotte's experiments and the views of Hume," T.R. Miles
has critically examined Michotte's work from a linguistic point of view,
and his criticisms if valid would equally apply to Whitehead's observations.
Miles points out that nowadays in contrast to Hume's time we distinguish
philosophical questions from psychological ones. As an example of the
former, he gives "Can sentences containing the word 'cause' be replaced
without loss of meaning by sentences containing the words 'if ... then' ?"[10]
This question, he tells us, is a conceptual one and cannot be settled by
empirical tests, although empirical evidence may be relevant to the conclu-
sions reached. Further, he goes on, although Hume is psychologically incorrect in denying that a causal impression in Michotte's sense occurs, a
Humean might still argue that all we see are two separate movements.

Miles says that the reason why a Humean would talk in this way is that
" 'I can see an X' entails in ordinary speech that an X is really there, and
since all that is really there is two movements it is impossible as a matter of
logic that we should see anything else."[11] We could even say, he continues,
that phenomenologically a causal impression sometimes occurs, whilst re-
fusing to say that we "actually see" causality. Miles does believe, however,
that a case could be made out conceptually for treating the word "causality" as a suitable accusative for the verb "see." Thus, he concludes, pro-
vided the phenomenology of the situation is not in dispute "it is hard to see
what useful purpose is achieved by discussing whether or not causality is
"really there" in addition to the two movements."[12]

On such a view a sharp distinction is drawn between philosophical questions and psychological ones: the former are assumed to be purely conceptual, the latter empirical. But if we examine the question whether a causal
statement can be translated into an "*if ... then*" one, a reference to an
empirical context comes in as soon as we specify what sort of causal situation
we have in mind. And if we concern ourselves with perceptual causality as
described by Michotte and Whitehead, this sort of translation cannot be
performed. We deal with more than a relationship of material implication.
The cause is linked with the effect in a changing perceptual structure. As
Miles himself points out, the translation of a causal statement into an
"*if ... then*" statement is only justified if we are not interested in perceptual

[9] *A Study of Psychology in Autobiography*, Vol. III.
[10] *The Perception of Causality*, p. 410.
[11] *Ibid.*, p. 413.
[12] *Ibid.*, p. 413.

causation. However, if such a conceptual formulation of causation is to have some applicability to our experience of causal activities, some reference is required to physical and psychological situations. An extensive knowledge of physical phenomena as well as theory is required, before we can proceed to formulate the concept of causality as it is used in physics.

The question which needs to be asked is, granted that causality is actually conceived as a Gestalt quality, are we justified in attributing to it an ontological significance? Thus though the phenomenology of the situation may not be in dispute, its ontology may be. Michotte, and for that matter Whitehead, regard these causal impressions as direct uninterpreted facts of experience. A Humean, on the other hand, would consider them to be the consequence of a learned response. If we take as self-evident the statement that only the two movements are really there (presumably in the physical world), then starting from such a premise it would, as long as we keep to logic alone, be formally impossible for us to see anything else. However, for such a statement to have empirical meaning we need to go beyond the conceptual formulation in terms of material implication, to the actual causal activities themselves. And one must remember that "material implication" is only one conceptual model of causality, a model which Kant, for example, would have found unacceptable, conceiving as he did causality as an *a priori* necessary relation.

In any case traditional philosophical accounts of causation do make certain factual assumptions. A good example of this is Hume's appeal to experience when he attacks the view that causation is a necessary relation. Implicit in this appeal is the assumption that in the physical world events are distinct, and related to each other after the fashion of balls drawn from an urn. On such a combinatorial model there is no intrinsic reason why one event should be followed by any other. To explain why despite this we nevertheless have direct awareness of causal efficacy, Hume, as Whitehead remarks, had to introduce a feeling of expectancy based on past associations. However, it does not seem to be the case that the habit of expectancy is alone at the root of our perception of causality.

It is true that Michotte's and Whitehead's accounts of perceptual causality, may not help us to understand how the concept of causality is used in contemporary epistemology. Their discussions, however, seem to have more of an ontological than an epistemological import. They are of some importance if they are taken as illuminating our primitive perceptual experiences, in which our more sophisticated notions of causality presumably have their root. Although if we try to see what sort of relationship there is between perceptual causality and the causality discussed in physical theory, epistemological questions do arise.

It must be remembered, however, that Michotte's experiments were carried out under laboratory conditions and hence have a somewhat artifical character, since physically they consist of separate movements of two distinct objects, which, nevertheless, give us the impression that one object is influencing the other. It would not necessarily follow that the processes in the physical world, would be independent of each other in the same sort of way: because of this "perceptual causality" may be some guide to physical causality at least on a macroscopic level.[13]

One criticism of Michotte's and Whitehead's positions can be made; despite their rejection of the view that interpretation enters into our causal impressions, this may yet occur but in a completely unconscious way. Although our causal categories may have, as it were, some deeply rooted instinctive basis, we cannot completely neglect the part played by past experience and learning in our recognition of causal conditions. Some sort of causal scheme in terms of which we come to see things is probably built up from childhood onwards, and may snap into action by giving us an impression of causality when we find ourselves confronted by one of Michotte's laboratory experiments. As adults it is doubtful whether our perceptions are entirely free of some sort of conceptual interpretation. Perhaps the truth after all lies closer to Kant than to Whitehead, Michotte or Hume.

---

[13] If it were not it is hard to see how man at least could have survived biologically.

CHAPTER 13

RELIGION, DEITY AND THE ORDER OF NATURE

I. RELIGION

There have been various accounts and interpretations of Whitehead's natural theology. It has been assumed, for example, by some writers that he is giving a new kind of ontological argument for the existence of God in terms of the necessary elements disclosed in experience. In discussing Whitehead's natural theology one thing at least is clear, that he is basically concerned with pointing out those general features of order we find in the universe, which he believes gives rise to our particular religious ideas, emotions and forms of behaviour. His aim, he says, "was to give a concise analysis of the various factors in human nature which go to form a religion ... and more especially to direct attention to the foundation of religion on our apprehension of those permanent elements by reason of which there is a stable order in the world, permanent elements apart from which there could be no changing world."[1]

Religion for Whitehead is then founded on our apprehension of certain factors of permanence (or order) in the world. But he also recognises the subjectivity of the particular religious emotions, the social character of the associated forms of behaviour, and the dependency of our religious picture of the world on the culture and period we live in. As he puts it, the principles of religion may be eternal but the expression of these principles is usually in terms of the imaginative picture of the world entertained in particular ages and among particular peoples.

Whitehead makes the same point in another way when he says: "Religion should connect the rational generality of philosophy with the emotions and purposes springing out of existence in a particular society, in a particular epoch ... Religion is an ultimate craving to infuse into the insistent particularity of emotion that non-temporal generality which primarily belongs to

---

[1] *Religion in the Making*, Preface.

conceptual thought alone."[2] Whitehead is once again drawing our attention to what he takes to be the most general experience upon which religion is based, namely, our experience of permanence or eternality amid change. This ultimate feeling, he tells us, is expressed in the first two lines of the well known hymn

> Abide with me;
> Fast falls the eventide.

Here, he says, we find formulated the complete problem of metaphysics: "the first line expresses the permanences 'abide,' 'me' and the 'Being' addressed; and the second line sets these permanences amind the inescapable flux."[3]

Although the precise form our religious experience takes will vary with the religion we practise and with the particular epoch and society in which we find ourselves, Whitehead would claim that all religious experience is centred round the same general notion of order (or permanence) with its suggestion of some kind of necessity or law in nature. He points out that from the belief that the universe was subject to divine law sprang the unquestionable faith in the reign of Natural Law. "Millions of men have marched to battle fiercely nerved by intense faith in law imposed by the will of inflexible Allah, Law sharing out to each human his inevitable fate."[4] Whitehead further makes the point in *SMW* that the Greek belief in the inevitability of fate and their attitude of rationality was transmitted through the abstract principles of Roman Law to medieval thought. And that in its turn the medieval church impressed on Europe the notion of the detailed providence of God, which was one of the factors by which the trust in the order of nature arose.[5]

Whitehead is therefore led to ask: "What is the status of the enduring stability of the order of nature?"[6] There is, he says, the summary answer which refers nature to some greater reality standing beyond it, which has in the history of thought taken on different names, "Jehovah, Allah, Brahma, Father in Heaven, Order of Heaven, First Cause, Supreme Being, Chance."[7] But Whitehead does not want to proceed from our conceptual recognition that there is such an order of nature to the belief that this is due to some supernatural being tyrant-like imposing it. The task he sets himself is to

[2] *PR*, p. 21.
[3] *PR*, p. 296.
[4] *AI*, p. 173.
[5] *SMW*, pp. 22-24.
[6] *SMW*, p. 115.
[7] *SMW*, p. 115.

discover whether nature does not in its very being show itself as self-explana-tory, and whether this order does not arise from the relations between things themselves.

II. DEITY

Broad has said with *PR* in mind that "It is often desperately difficult to un-derstand what it is that Whitehead is asserting. When one is fairly sure of this, it is often equally hard to discover what he considers to be the reason for asserting it; for he seems to be "not *arguing* but just *telling* you."[8] This tendency is nowhere more apparent than in Whitehead's discussion of God in *PR*. It is not possible to give here a detailed analysis of the complex-ities of his position as it is developed in his chapter on "God and the World" in *PR*.[9] Instead, we shall content ourselves with some account of what we take to be the main points he wishes to make there.

We have already seen that Whitehead is concerned in his account of the concept of God's functioning in the universe, with indicating the elements of permanence or eternality in the world, with which, he claims, religion has essentially concerned itself. Thus he tells us, "The theme of Cosmology, which is the basis of all religions, is the story of the dynamic effort of the World passing into everlasting unity, and of the static majesty of God's vision, accomplishing its purpose of completion by absorption of the World's multiplicity of effort."[10] What Whitehead seems to be saying in this rather abstruse passage is that in cosmology we are concerned with the way the creative activity of the world (or course of events) obtains a definite form of order, and how this form of order which in itself is purely potential is given a qualitative content by the course of events. He is claiming that this process of the intermingling of change and permanence has been intuitively singled out by religion as a basis for its teachings.

Whitehead makes the same point in another way, when he says: "God is dipolar. He has a primordial nature and a consequent nature ... The primordial nature is conceptual, the consequent nature is the weaving of God's physical feelings upon his primordial concepts."[11] Whitehead is referring here to the two aspects of his concept of God: (a) its formal con-ceptual aspect where it is entirely unlimited in its range of possibilities, and (b) its physical aspect as relating the multiplicity of events in our everyday

[8] C.D. Broad, "Alfred North Whitehead," p. 144.
[9] *PR*, Part IV, Chap. II, pp. 484-497.
[10] *PR*, p. 494.
[11] *PR*, p. 488.

world. In (b) the formal relations are, as it were, interwoven with the individual particularities of events.

Whitehead describes the "primordial nature of God" as "the unconditioned actuality of conceptual feeling at the base of things."[12] But he also believes that God's conceptual feelings are empty of all content, and this would seem to imply that the primordial nature consists of a system of possibilities. More light is thrown on this concept when Whitehead says: "eternal objects as in God's primordial nature constitute the Platonic world of ideas."[13] However, as these eternal objects are devoid of any specific exemplification, Whitehead would seem only to be referring here to their "relational essence" or relations of order. He has indeed told us that the order of nature prevalent in our cosmic epoch "exhibits itself as a morphological scheme involving eternal objects of the objective species."[14] Such eternal objects are also described by him as "the mathematical platonic forms."[15] Whitehead may then be said to regard the primordial nature of God as the ordering element in nature, which relates the multiplicity of events into an ordered system. In this way it determines general conditions to which all events have to conform.

From this point of view Whitehead tells us, God may be considered as the principle of concretion or limitation,[16] since it limits the course of events to being ordered in this manner, and thereby acts as a ground for induction. If our predictions are to be in any way justified, then underlying all our arguments must be the presupposition that the order of nature will continue in the future – that future events will be ordered in the same way as the events in the present.

Whitehead's dipolar conception of God needs to be looked at against the background of the different cosmologies which have been elaborated in the past in terms of one or other of the two opposing principles of "change" and "permanence." An early example of the former principle is to be seen in Heraclitus' saying that "All things flow" and of the latter in the doctrines of Parmenides and Zeno who emphasised the permanence of things. In more recent times empiricists have stressed the fleeting nature of experience, whilst the rationalists have concentrated on the universal categories disclosed by reason as exemplified in things. But as Whitehead sees it, an understanding of the universe requires that each such principle exhibit the impress of

---

[12] *PR*, p. 486.
[13] *PR*, p. 63.
[14] *PR*, p. 414.
[15] *PR*, p. 413.
[16] *SMW*, p. 208 and *PR*, p. 488.

the other. "There is not," he says, "the mere problem of fluency *and* permanence. There is the double problem ... The first half of the problem concerns the completion of God's primordial nature by the derivation of his consequent nature from the temporal world. The second half of the problem concerns the completion of each fluent actual occasion by its function of objective immortality, devoid of 'perpetual perishing.' "[17]

On Whitehead's view then if we consider the universe from the point of view of either of these two principles, we merely restrict ourselves to certain limited aspects of our experience of nature. There is rather a dual process in which the scheme of order is given a qualitative temporal content, whilst the qualitative elements (or events) are given a formal structure. Thus each event (or occasion) is experienced as related to other events in one spatio-temporal system made up of societies of events (or physical objects). As a member of such a society an event is devoid of "perpetual perishing," since a society unlike an event, which only becomes and perishes, has the peculiar character of "objective immortality." It therefore enjoys a history expressing its changing relations to changing circumstances.

Whitehead's account of the interplay of God and the World seems to be a particular example of Kant's dictum "concepts without intuitions are empty, intuitions without concepts are blind," since he is essentially concerned with the basic interrelationship of structure and content in our experience of nature. Furthermore, Whitehead would seem to be restating here Kant's question. "How is science (or for that matter everyday experience) possible?" But he does this from a realist rather than an idealist position. As we have seen Whitehead does not believe that we would be able to make rational and coherent statements about nature and ourselves unless our experience exhibited some very general principles of order.

III. THE ORDER OF NATURE

Another more formal way in which Whitehead expresses the order implicit in nature in *PR*, is in terms of his concept of the "Extensive Continuum," which he pictures as one relational complex (of extensive relations) underlying the world, past, present and future.[18] Some Whitehead commentators would be unhappy about any attempt to show that the primordial nature of God and the Extensive Continuum have certain properties in common.

---

[17] *PR*, p. 491.
[18] *PR*, cf. p. 91.

However, in a number of ways they perform a similar function in his system, as far as the determination of the order of nature is concerned. It is true that these concepts occur in *PR* in different contexts: the former in Whitehead's discussion of natural theology (particularly in Part V), the latter in an account of the most general properties of order any cosmic epoch needs to exemplify (Part II, Chap. II and Part IV). In his discussion of the Extensive Continuum, Whitehead would, however, seem to be expressing on a purely formal level aspects of our experience of an order of nature, which have been described more concretely in terms of religious experience.

What Whitehead means by the Extensive Continuum can be seen from his description of it as "a complex of entities united by the various allied relationships of whole to part, and of overlapping so as to possess common parts, and of contact, and of other relationships derived from these primary relationships."[19] The notion of a continuum, he tells us, "involves both the property of indefinite divisibility and the property of unbounded extension. There are always entities beyond entities."[20] Although he would seem to be concerned here with a purely formal concept, he does consider the Extensive Continuum as a real element in our experience. It expresses, he says, a fact derived from the world and concerning the contemporary world. "The reality of the future is bound up with the reality of this continuum."[21] Whitehead would seem to assume that we become aware of this continuum in a rough and ready way in our experience, and then by generalisation arrive at the conception of an abstract formal system which enables us, as it were, to extrapolate to the structure of future events. As he puts it: "In this sense, it is Kant's 'form of intuition': but it is derived from the actual world *qua datum*, and thus is not 'pure' in Kant's sense of that term."[22]

As an abstract system of part and whole relations the Extensive Continuum does bear a certain resemblance to the mathematical continuum. It however differs from this continuum conceived as a closely packed infinite series of points (i.e. possessing infinite divisibility), since Whitehead regards points as derivative from what he takes to be the more general extensive relationships. Whitehead's conception of an Extensive Continuum, as a connected system of abstract regions rather than an infinite series of discrete points, brings his view closer to the intuitionist's continuum, where the fundamental relation is that of part and whole rather than element to set. Whitehead may then be said to be assuming something like a general prin-

---

[19] *PR*, cf. p. 91.
[20] *PR*, p. 91.
[21] *PR*, p. 92.
[22] *PR*, p. 100.

ciple which is found not only in our direct experience of nature, but is also exemplified in our abstract mathematical conceptions. This general principle of order would be identified by Whitehead with what he terms metaphysical necessity, since he assumes that it will be exemplified in every cosmic epoch.

It may seem a little suprising that in what is primarily a discussion of Religion and Deity, one has to introduce a discussion of the general nature of order, and in particular an account of Whitehead's concept of the Extensive Continuum. This would seem to be another example of a tendency to be found in Whitehead's writings, namely his attempt to relate the purely mathematical concepts of pattern and order with topics of vital human interest, in this case religion. This tendency is particularly evident in his much quoted statement that in time symbolic logic – the examination of pattern with real variables – would become not only the foundation of aesthetics but also of ethics and theology.[23] Schilpp has, for example, remarked of Whitehead's attempt to connect mathematics with morality, that he remains unconvinced by Whitehead's "efforts to get both mathematics and the Good, so to speak, out of the same pail." He goes on: "I fear that Mr. Whitehead's philosophical thought has become so saturated by ... his life-long interest in mathematics – instead of with Spinoza seeing all things *sub specie aeternitatis* or with the naturalists' *sub specie naturae* – Mr. Whitehead simply cannot help seeing all things *sub specie mathematicae*."[24]

Although in Whitehead's discussion of religion and Deity analogies may be found with the more formal notions he develops elsewhere in *PR*, this does not mean that he essentially regards religious concepts *sub specie mathematicae*. What he is saying is that the basic experience from which religion arises – our awareness of permanence amid the flux of events – is reflected in the particular thoughts, emotions and purposes which go to form our whole religious way of life. In logic and mathematics these general conceptions of pattern and order have been taken up and treated on a more abstract level. Whitehead is not therefore claiming that religion is reducible to mathematical pattern. If anything the position is the other way round: the formal concept of order only deals with certain abstract features of the more concrete data with which religion is concerned. Historically, at least, the concept of order was first systematically depicted in man's religious picture of the world.

    [23] "Remarks," *Philosophical Review*, 1937, p. 186.
    [24] P.A. Schilpp, "Whitehead's Moral Philosophy," in *The Philosophy of Alfred North Whitehead*, p. 613.

BIBLIOGRAPHY

Bibliography of the main works of Whitehead (excluding articles and reviews) giving date of first publication. Where the work is published both in England and America, the American publisher and publication date is given in parentheses.

1. *A Treatise on Universal Algebra, with Applications*, Vol. I (only vol. published), Cambridge University Press, 1898.
2. *The Axioms of Projective Geometry*, Cambridge University Press, 1906.
3. *The Axioms of Descriptive Geometry*, Cambridge University Press, 1907.
4. (With Bertrand Russell), *Principia Mathematica*, Vols., I-III, Cambridge University Press, 1910-1913.
5. *An Introduction to Mathematics*, Williams and Norgate, 1911 (Henry Holt and Co., 1911).
6. *The Organisation of Thought, Educational and Scientific*, Williams and Norgate, 1917 (J.B. Lippincott and Co., 1917).
7. *An Enquiry Concerning the Principles of Natural Knowledge*, Cambridge University Press, 1919.
8. *The Concept of Nature* (Tarner Lectures, 1919), Cambridge University Press, 1920 (The Macmillan Co., 1929).
9. *The Principle of Relativity, with Applications to Physical Science*, Cambridge University Press, 1922.
10. *Science and the Modern World* (Lowell Institute Lectures, 1925), Cambridge University Press, 1926 (The Macmillan Co., 1925).
11. *Religion in the Making* (Lowell Institute Lectures, 1926), Cambridge University Press, 1926 (The Macmillan Co., 1926)
12. *Symbolism, Its Meaning and Effect* (Barbour-Page Lectures, 1927). Cambridge University Press, 1928 (The Macmillan Co., 1927).
13. *Process and Reality. An Essay in Cosmology* (Gifford Lectures, Edinburgh, 1927-28), Cambridge University Press, 1929.
14. *The Function of Reason* (Louis Clark Vanuxem Foundation Lectures, Princeton, 1929), Princeton University Press, 1929.
15. *The Aims of Education and Other Essays*, Williams and Norgate, 1929 (The Macmillan Co., 1929).

16. *Adventures of Ideas*, Cambridge University Press, 1933 (The Macmillan Co., 1933).
17. *Nature and Life*, Cambridge University Press, 1934 (University of Chicago Press, 1934).
18. *Modes of Thought*, Cambridge University Press, 1938 (The Macmillan Co., 1938).
19. *Essays in Science and Philosophy*, Rider, 1948 (The Philosophy Library, 1947).

NAME INDEX

SUBJECT INDEX